Floodgates

Susan Ingram
Markus Reisenleitner
Cornelia Szabó-Knotik
(eds.)

Floodgates

Technologies, Cultural (Ex)Change and the Persistence of Place

PETER LANG
Frankfurt am Main · Berlin · Bern · Bruxelles · New York · Oxford · Wien

Bibliographic Information published by Die Deutsche Bibliothek
Die Deutsche Bibliothek lists this publication in the Deutsche Nationalbibliografie; detailed bibliographic data is available in the internet at <http://dnb.ddb.de>.

ISBN 3-631-53988-6
US-ISBN 0-8204-7730-3

www.peterlang.de

Table of Contents

Introduction

SUSAN INGRAM
MARKUS REISENLEITNER
CORNELIA SZABÓ-KNOTIK

Movement is not just the experience of shifting from place to place, it is also linked to our ability to imagine an alternative. (Papastergiadis 11)

This edited volume is based on papers presented at Ryerson University in Toronto on 5-8 August 2004, the fourth in a series of international workshops at which an interdisciplinary group of scholars from Central Europe, North America, and Asia have met to explore how the imaginary geo-cultural space of »Central Europe« has acquired meanings over the past two centuries, and how traveling images of an »othered« cultural space – inserted into specific regional, national and social contexts and appropriated for negotiations of cultural identity and belonging as well as exclusion and colonization – have laid the basis for a cultural essentialism which thinks culture through space and negotiates cultural status through de-historicized notions of place and territory.[1] In previous workshops we have focused on modernity and its cultural practices, which have more often than not proven to be adept at traveling, and we have explored how such traveling has led to different modes of insertion and the re-contextualization of those practices, often for implicit or explicit political purposes of identity formation and modes of exclusion.

This workshop focused on the specific factors and forms of mediation that mobilize, inhibit or prevent culture/s from traveling. More specifically, we asked: what are the technologies, media, and forms of representation that render im/possible cultural ex/change? As in our previous volumes, the case studies collected here are intended as probes into a wide variety of contexts. Approaches to the highly complex processes delineating cultural mobility, they interrogate the instability of cultural flows and attempt to determine how these flows are re-territorialized into stable spatialized forms of identification and imaginings of culture. The volume begins by exploring the centrality of technologies in globalizing circuits of culture; examples are taken from what have historically been the central technologies of diffusion, radio and television, as well as a less obvious medium, the piece of textile commonly described as a t-shirt, which, as is argued, has become a versatile and practically paradigmatic global message board. The focus of investigation then shifts to the question of the centrality of »Central« Europe in establishing the distinction of European culture globally, with distinction understood territorially as well as socially. The first part of this probe examines prominent cultural products »made in Austria« – the Vienna Philharmonic, *neue Volxmusik* and a Schubert biopic – that have exploited, and in the case of Schubert queered, the status of the Land of Music's cultural heritage to propel themselves into the orbit of the global, while a second subsection delves into a different strategy, that of persistence. Taking their cues from, among others, Vivian Sobchack's approach to cinema history and Svetlana Boym's work on nostalgia, these contributions tackle the sometimes romantic, sometimes surreal knots that bind or blind places, specifically Berlin, L'viv and the Balkans, to specific histories and memories, and thus set off equally binding and blinding processes of identity formation. The final probe then returns to the global and probes in how far centrality has been achieved by marking music and filmic imaginaries as »European« or simply »Western.«

What struck us about the contributions as they came together is the way in which a focus on technologies consistently directed attention towards some media for cultural production and away from others. While it was not a deliberate choice on our part – contributors were encouraged to explore the same wide variety of cultural phenomena they have in the past –, case studies no longer dealt with works of literature or world fairs but rather concentrated on film and music. New technologies seem to have begun making it possible for sound and vision in particular to travel in ways not previously possible and, in doing so, they seem to be influencing the modes in which communities are imagined in far more complex ways than print capital did. Once the imagining of communities is carried out via new technologies such as CDs, DVDs and the internet, global circulation becomes a reality inherent to the form of mediation and resonant with the possibilities of community formation beyond territorial borders. This problematizes territorial signifiers of modern identity formation such as »nation« and and forces them to reassert themselves in interesting new ways.

It may be, as Hal Foster pointed out in reviewing John Seabrook's *Nobrow: The Culture of Marketing, the Marketing of Culture*, that we have simply reached a new phase (or perhaps more appropriate to gaming culture: level) of the culture industry:

> One could identify three phases of this industry in the 20th century: the first in the 1920s as radio spread, sound was connected to film and mechanical reproduction became pervasive (Debord identified this as the formation of ›spectacle‹); the second in the postwar perfection of a consumer capitalism (with its image world luridly exposed by Warhol); and the third, perhaps, in the digital revolution and dot.capitalism of today.

The global culture industry may have become a seamless semiotic reservoir, into which cultural producers can dip at will – Austrian musicians as well as Japanese film directors, Ukrainian intellectuals as well as American TV news producers –, yet no unified image of that industry emerges from the following pages, any more than a unified image of »Central Europe« or »North America« emerged in our previous volumes. Rather, the more seamless it becomes, the more paramount attention to historical specificity becomes as well. Just as our previous volumes have clarified how motion through space – whether physical or imaginary – has helped shape, crystallize and negotiate images of the cultural other in contact or transit zones, where people, images and cultures meet in asymmetrical relations of exchange, domination and subordination, and where tourists, exiles, travelers, displaced commodities and »foreign« cultural practices generate powerful, as well as (potentially) subversive, visions and imaginings, here the multidimensionality of motion is foregrounded. Traversing space and time together reveal them to be mutually imbricated processes that open up avenues and offer insights into what otherwise might be written off as banal or quirky. Historical location matters. As is shown repeatedly in the contributions, it is precisely in attending to such locationality that inroads into the multiple layers (or again: levels) of complex artistic imaginaries are opened up for analysis, a kind of analysis that sees in the manifold varieties of nostalgia a wily boxing opponent and seeks to land a knock-out punch which would release the containment of place-bound meanings and identifications and allow for floods of possible exchanges and resignifications. Given the unsettling, variegated paths global academics are increasingly forced to ply, it is perhaps to be expected that some would develop well-honed built-in motion detectors. That such productive exchanges continue among our contributors reflects not only on their eagerness to engage – whether with new technologies or new ideas – but also on their openness and flexibility in the face of the daunting demands that globality entails.

We are once again grateful to everyone who helped us put together this volume and the workshop on which it is based, especially James Deaville for the wonderful occasion of fruitful discussions and stimulating exchange in Toronto.

NOTES

1 The workshops have resulted in three volumes edited by Susan Ingram, Markus Reisenleitner, and Cornelia Szabo-Knotik: *Identität. Kultur. Raum: Kulturelle Praktiken und die Ausbildung von Imagined Communities in Nordamerika und Zentraleuropa* was published by Turia + Kant in 2001; *Reverberations: Representations of Modernity, Tradition and Cultural Value in-between Central Europe and North America* and *Ports of Call: Central European and North American Culture/s in Motion* appeared with Peter Lang in 2002 and 2003.

REFERENCES

Boym, Svetlana. The Future of Nostalgia. New York: Basic Books, 2002.

Foster, Hal. »Slumming with Rappers at the Roxy.« *London Review of Books* 22.18 (21 September 2000).

Papastergiadis, N. *The Turbulence of Migration: Globalization, Deterritorialization and Hybridity*. Cambridge: Polity Press, 2000.

Sobchack, Vivian, ed. *The Persistence of History: Cinema, Television, and the Modern Event.* New York: Routledge, 1996.

Probe I:

The Centrality of Technologies in Globalizing the Circulation of Culture

»Something Extra for the Boys«: »Radio Girl Friends,« the BBC, and Forces Broadcasting, 1940-1944[1]

CHRISTINA BAADE

Anne Shelton we love you
Yes, indeed we do.
We of the 8th Army.
You thrill us through and through.
Only the other evening
We were listening to your songs
We of the 8th Army as we roll along.
Cheerio Ann, we hope to hear from you
And if we are lucky on the air too.

»Anne Shelton in the Desert,« by Private Randall, 153 Infantry Brigade, North Africa[2]

Though perhaps not the finest example of English verse, this poem demonstrates the strong convictions held by listeners in World War II Britain about popular song and the voices of female radio crooners. Not all were as enthusiastic as Randall. A counter-example to the devoted soldier-listener's lyrics, which he sent to the popular young singer in April 1943, are those of AP Herbert's satiricial »Ode to a Crooner,« in which the woman singer is called a »whining witch.« Private Randall and his mates listen hopefully for their favorite singer on the radio while Herbert acts decisively to gag the crooner by turning off his wireless set after giving her a good scolding. Their verses delineate not only striking differences in public opinion, but also the BBC's fraught approach to popular song and women's voices in its war effort.

During the Second World War, the British Broadcasting Corporation embraced bright radio entertainment as a means to help relieve stress, bring cheer, and provide escape for civilians and members of the Forces. Just as enlightening talks, challenging classical music, and serious drama under Sir John Reith's director-generalship had been justified as morally improving and culturally uplifting, variety and dance music were now justified as morale building. The moments of pleasure and relaxation that radio could provide aided the public at a time when, as the BBC publication *The Listener* put it, »to be of good cheer becomes more than a duty to oneself: it becomes a duty to one's neighbor« – and, by extension, to the nation.[3]

Unlike more uplifting material, however, the broadcasting of popular dance music was not justified by its inherent qualities. Rather, it was valuable in morale maintenance only so long as it entertained, relaxed, and cheered most of the public. The findings of the BBC's Listener Research Department, founded in 1936 by pioneering pollster Robert Silvey, played an important role in determining and justifying the schedule, quantity, and finance of entertainment programming throughout the war. Tracking majority preferences represented only the surface of Silvey's concerns: he and his staff worked to understand specific audiences, especially those involved in national service. The most important of these special groups were the men serving in the Forces. Servicemen presented the BBC with two major challenges: they lived outside its traditional address of the domestic sphere and their preferences were notoriously difficult to track. Thus, in public – and even in internal BBC – debate, their tastes and listening needs were evoked to assert competing ideologies of gender, nation, and war.

One of the most influential figures in determining how best to broadcast to Forces stationed abroad was Cecil Madden, the director of the Overseas Entertainments Unit, which was established in 1940 and extended to a twenty-four-hour service in 1942. Located in central London at the underground Criterium Theatre, where stalls and boxes provided a bomb shelter and dormitory for performers who broadcast throughout the night, the Unit sent variety and dance music around the world. In his broadcasts to the Forces, Madden made unprecedented use of women producers, announcers, and singers. While their work could be narrated as a »Rosie the Riveter« story of women on the home front, their contemporaries praised their unique abilities to represent the values of home and homeland to men stationed abroad. As the populist *Daily Mirror* opined, »The women of Britain owe a debt to these girls who, day by day, keep thoughts of home in the minds of a great army of men. [...] Their names are known to every soldier who fights for us overseas, but England – strangely enough – has never heard of them.«[4]

Two years later, »England« had. In February 1944, the BBC replaced the domestic Forces Programme with the General Forces Programme. The change made Overseas Entertainments Unit productions available at home: outspoken members of the public greeted the women announcers and singers with outrage, as inappropriate to virile fighting men. To contextualize their outrage, I will sketch the BBC's approach to Forces programming, examine issues of sentimentality in dance music, and explore the work of three women active with the Overseas Entertainments Unit: the singers Vera Lynn and Anne Shelton and the producer Joan Gilbert.

Although the BBC's ideal prewar listener may have been a man in his easy chair, surrounded by family, attentively enjoying a concert, drama, or even a variety show after a long day, it soon recognized that servicemen lived in circumstances far removed from the domestic comforts of the radio hearth (to use Simon Frith's term).[5] The BBC instituted the Forces Programme early in 1940 to serve the special listening needs of men in the British Expeditionary Forces stationed in France during the first winter of the war. The Forces Programme played popular music extensively and represented the BBC's first acknowledgement that inattentive listening was something that could be catered to, rather than corrected. It also diverted troops from listening to Continental and even enemy broadcasts, and it provided civilian listeners with a much-needed alternative to the Home Service.

The Forces Programme represented much more than a distraction from enemy radio. It was special programming, created after careful listener research, directed to an audience that had arisen as a direct result of the war.[6] Soldiers comprised a new listening demographic: as the *Listener* asserted, »The listeners in the Forces are not new listeners, but they are a special section of the old listening public transferred to new and peculiar surroundings. Bodies of men need an approach different from that most suitable for the family circle.«[7] Although less than a year before Listener Research had found that gender accounted for few differences in listener tastes, the dislocations of war ended the BBC's single-minded address to the radio hearth.[8] The Forces Programme avoided programming that required concentrated listening, such as long talks and symphony concerts. Instead, BBC programmers envisioned the troops listening in groups in canteens and barracks to programmes »that do not suffer unduly by interruption, either by conversation or by the call of duty.«[9] In addition to news, sports, and humor, the programme featured dance bands, theatre organs, variety orchestras, and military bands.[10]

An important function of the Forces Programme was to link soldiers stationed in France (and in remote camps and billets throughout Britain) with their families back home.[11] One of the most direct techniques was the musical »messages programme,« a format introduced in February 1940 by the BBC's theatre organist, Sandy Macpherson.[12] In essence, it was a request programme: servicemen in France as well as their mothers, wives, or sweethearts could write Sandy with a tune request and a message for a loved one; the recipient would be notified before the programme aired. To further the personal touch, Macpherson began to invite some of »the womenfolk« and children into the studio to deliver their messages themselves.[13] Other programmes evoked homey pleasures more directly. In 1942, the BBC created *At Home to the Forces* especially for communal listening; it featured diverse, if somewhat middlebrow, bookings, from the BBC Symphony Orchestra to the popular vocalist Vera Lynn and variety »of the more subtle kind.« The programme complemented an army organization that linked soldiers to civilian homes, where they could »take their boots off and stretch their legs in easy chairs [...] and enjoy all the other pleasant amenities of decent family life« – including the radio, »as listening has now become one of [its] essential ingredients.«[14]

The Forces Programme was thus intended for groups of men listening outside the domestic circle: it both appealed to their special circumstances and reminded them of their homes and families. It is ironic that, despite the BBC's gendered rhetoric about »bodies of men,« more civilians than soldiers listened to the programme; its audience grew to be fifty percent larger than that for the Home Service.[15] BBC Listener Research found that working class listeners favored the programme overwhelmingly, as did middle class listeners the younger they were.[16] Meanwhile, Mass-Observation, an independent social research body, found that working women also preferred it to the Home Service.[17] The qualities that made the Forces Programme ideal for fighting men also appealed to civilians, rendering it an effective alternative programme. By 1942, most servicemen were stationed either too far abroad to hear the service or, being stationed in Great Britain, could also access the Home Service.[18] Mass-Observation asserted, »The Forces nowadays is really the housewives' programme, and might well be renamed. It is tap listening for the home.«[19]

Whatever the realities of its listenership, the Forces Programme continued to symbolize what the troops heard. In 1942, faced with military setbacks, critics in Parliament and the press began to worry about the »sentimental and sloppy muck that goes on hour after hour on the Forces' programme« and the effects it would have upon the men fighting overseas.[20] A Lieutenant Colonel who had served in the previous war claimed, »The sickly and maudlin programmes are largely responsible for the half-hearted attitude of so many people towards the war and its seriousness generally.«[21] Despite the concerns with flabby broadcasting in the Forces Programme, BBC Listener Research could detect no matching frustration on the part of actual servicemen.[22] Lacking empirical data, a columnist in the dance music weekly, *Melody Maker,* compared the effect of the BBC's output of sentimental dance music to that of »an overdose of a sedative drug« and the public's craving for sentimentality to the »psychological demand for sugar« brought on by the war.[23]

The BBC agreed with the assessment. During 1942, the Corporation embarked on a sweeping re-evaluation of its dance music programming. Arthur Bliss, the BBC's new Director of Music, wanted to treat men in the forces like »princes« by providing them with the best possible music.[24] In the BBC Music Policy of 1942, he extended Music Department oversight to popular music administered by the Variety Department. For Bliss, classical music was an

»ultimate value, indeed a justification of life« while other music could »enrich leisure hours with entertainment.«[25] Popular music, however, was riddled with problems and required close supervision to curb its excesses. In July, the BBC banned from broadcast male crooners, overly sentimental and insincere female singers, and slushy songs. Until the end of the war (and beyond), the Dance Music Policy Committee met to vet songs and then singers to enforce a more virile British dance music. Cecil Madden observed in hindsight, »The flabby type of songs being written [...] w[as] having a drugging effect instead of making for a fighting spirit.«[26]

Madden and his Overseas Entertainments Unit helped lead the BBC's efforts to become »more in tune with the times«; they eliminated the most egregious crooning and »enervating« songs while encouraging more robust music.[27] His reforms encompassed issues of presentation style, use of singers, song lyrics, and dance bands. Nevertheless, because of his practical concerns with giving overseas forces the entertainment and music they wanted, Madden was a moderating force on the Dance Music Policy Committee. A case in point was his treatment of Vera Lynn, the popular »Sweetheart of the Forces.« While the BBC debated sentiment in music, her programme, *Sincerely Yours – Vera Lynn*, enjoyed a large, »remarkably steady« audience of twenty percent during two series in 1941 and 1942.[28] Lynn claimed never to have sung a song that she did not believe in. She performed in a heartfelt, intimate manner, and her »reassuring« voice with its distinctive »sob« was instantly recognizable.[29] Lynn, who influenced an entire school of sentimental, crooning imitators, was an exceptional figure for those worried about debilitating sentiment – I suspect she is the reason why the BBC specified that a female singer had to be both »overly sentimental« *and* »insincere« to be banned.

Madden located Vera Lynn's sentimentality in the »sloppy and rather insincere talking presentation« of *Sincerely Yours*.[30] The letter »to the men of the Forces [...] in words and music« featured Lynn reading messages that illustrated women's perseverance and fidelity on the home front and scripted platitudes about love.[31] The talk functioned as continuity patter between her singing of current popular songs and Tin Pan Alley favorites from the 1920s and 1930s, along with light instrumental arrangements featuring Fred Hartley and his Music. At the start of the first run of *Sincerely Yours*, Lynn announced every week the name of a lucky man in the forces who was now a father – preferably of twins or triplets – and delivered a bouquet to his wife the next day. The feature quickly succumbed to its own success: on Sundays, hospitals and district nurses flooded the BBC's telephone lines with »news of babies,« so the producer, Howard Thomas, discontinued the stunt.[32] In contrast, Madden offered Lynn and similar singers to the forces in more straightforward programmes, such as *Starlight*, a nightly fifteen-minute programme that featured singing but no speaking. Madden judged sentimentality necessary to maintain programme balance – as long as it was used with moderation. Lynn set the standard for female vocalists who broadcast to the troops. As Madden later reflected, »Her singing programmes were so popular, her tones so even and so reassuring to the Forces everywhere, it was inevitable she would be regarded as No. 1 Sweetheart, and she deserved it. Her beautiful voice quality was unmistakable, her fan mail immense.«[33]

In broadcasting for forces stationed abroad, the Overseas Entertainments Unit sought to embody the values of family, home, and nation. In the array of articles that accompanied the Unit's transformation to a twenty-four hour service in early 1942, a widely published Ministry of Information piece described the BBC as providing a comforting »life-line« between soldiers and their families in the home country.[34] Sandy Macpherson's messages programme provided the template for dozens of programmes, to which Madden added another innovation: all programmes were hosted by women, or »radio girl friends,« as the press put it.[35] This was a revolutionary change: before the war there were almost no women announcers at the BBC, and many considered female voices unsuitable for radio broadcasting. Madden employed nine women announcers and three female producers in his unit. Moreover, he favored female singers over male, necessitating special negotiations with dance bandleaders, whose practice was to employ several vocalists of each gender. He based his reasoning on what was appropriate to a nation at war:

> We invariably [...] cut out male »crooners« or »high tenors« and include bright girls, for whom there is a real demand by, say, Forces in the Desert. Male crooners are quite

> divorced from the reality of the times, and to plug the girls is to show the men that the women are doing men's jobs these days.[36]

The so-called croonettes who sang in messages programmes, in purely musical programmes like *Starlight* and in a growing number of series that featured individual singers, carried out BBC propaganda objectives with an intimate touch. In *Parade,* a weekly magazine for forces in the Middle East, a writer employed purple prose to describe how »the girls who sing from London […] bring us London in their song.«[37] The special programmes included *Calling the British Forces in Malta*, featuring the singer Anne Shelton and announcer Avis Scutt, *A Date in the Desert*, featuring singer Doreen Villiers, and *Appointment with Beryl*, featuring Beryl Davis.[38] Although the term »pinup girl« did not enter common usage until 1943, women announcers and singers functioned as what Madden later called »pin-up personalities.«[39] Their radio personas combined girl-next-door charm with non-threatening sex appeal, building up rather than challenging their listeners' masculinity.[40]

Anne Shelton was an outstanding example of the cohort of very young crooners who were promoted heavily during the war by both the BBC and the dance bandleaders under whose auspices they worked. Shelton, who was eleven when war was declared, won her first radio audition a year later and continued to broadcast regularly thereafter. To circumvent child labor laws, her age was inflated to fifteen, an action rendered convincing by her excellent intonation and astonishingly mature contralto voice. By 1941, she was featured in *Calling Malta*, where she sang requests and developed a reputation for versatility. She could sing sentimental ballads favored by singers like Vera Lynn along with comedy numbers and bright marching tunes.[41] By 1943, *Gramophone* described her as the best English crooner on the air.[42] Even before the critical acclaim, she became popular with the forces through the Malta programme and other overseas broadcasts, as thousands of letters attested. In August 1941, one captain wrote:

> Anne, Anne, Why must you do it? Here we are in the desert, war like, full of hate for the [H]un, ready for war and you come and sing in *Starlight*. That voice of yours, really,

> well I can't say what it does to one but everything else goes into the background. Do you look like you sound? If you do you must be delicious.[43]

Shelton's visual mystery was unveiled when a very mature headshot made the cover of an April 1942 *Parade* magazine, transforming her from mysterious voice to English pinup personality for British troops in the Middle East (a status helped by the caption that inflated her age to eighteen).

Madden selected Shelton to counteract what one soldier described as the »seductive and soothing, husky, intimate but mysteriously unattainable« voice of Lale Andersen.[44] Nazi-controlled Radio Belgrade broadcast her nostalgic »Lili Marleen« incessantly, which attracted a significant following among British troops. Responding to military concerns, the BBC decided to create a programme that could compete directly with Andersen's »Marleen.« Madden suggested that Walt Disney's »team of musical experts« be contracted to write a simple marching song, and continued, using language redolent of potential embodiment,

> If you really want to counteract the Marlene idea for sex-starved troops, give us a time every day at the same time and let us put on a Glamour programme with say Four girls… say a Blonde, a Brunette, and a Redhead – and an American girl too, for the USA troops in the Middle East. Plenty of songs, Records, all sorts of changes.[45]

His brainstorm for pinup girls on the air was soon pared down to a programme that would feature a single blonde, Anne Shelton, and an anglicized adaptation – »Lili Marlene« – for her signature tune. *Introducing Anne*, which debuted in October 1942, was developed by the producer David Miller to be broadcast every Sunday (as *Sincerely Yours –Vera Lynn* had been) to the Middle East and, simultaneously, over the Forces Programme at home. The Ambrose Players accompanied Shelton, who both sang and did the continuity talk. Alick Hayes, Shelton's producer from *Calling Malta*, was outraged when he heard the first broadcast: Miller was turning her into another sentimental Vera Lynn rather than utilizing Shelton's »natural tomboy personality.«[46] For over a year, Hayes had monitored and answered Shelton's fan mail and molded her persona for the

Malta programme in order to depict her, »not only as the possessor of a lovely voice, but as a cheerful and cheery personality: in fact as the sort of pleasant English girl whom a man serving overseas [...] would be glad to introduce to his wife or mother or sister or girl friend.«[47]

In contrast to the erotic danger of Anderson's voice, not to mention the risks of physical encounters with foreign women, the radio girl friends offered chaste intimacy, whether cheerful or sentimental, and represented the wholesomeness of British womanhood. While the Overseas Entertainments Unit might have made unprecedented use of women in its programming, the women did not necessarily control how their own voices, personas, or images were used. How Shelton felt about the way her persona was developed in *Introducing Anne* is not clear. However, in a 1994 interview shortly before her death, she related that when she had first heard the signature tune »Lili Marlene,« she had thought it awful but was told, »Well darling, it isn't whether you like it or not. You're being used to counteract the Nazi propaganda programme Goebbels is putting out.«[48]

The effect of the BBC's programming on troops stationed overseas was difficult to gauge quantitatively because Listener Research could only survey a statistically insignificant number of active duty troops. Nevertheless, letters to singers, announcers, and bandleaders bespoke their popularity, and performers formed strong opinions of the forces' tastes when they toured the Mediterranean and Middle East. One of the most useful documents on life and musical tastes in the forces was filed by Joan Gilbert, the »Sweetheart of the Rock,« who produced the messages programme *Hello Gibraltar* and visited her audience – to enthusiastic reception – in February 1943.[49] She visited hospitals, social functions, and numerous work sites, ranging from ack ack (anti-aircraft gun) installations to oil tanker crews. At an »Other Ranks« dance, Gilbert described dancing »two dances with about 40 different men, swapping partners each half round the room.« She reported on challenges to morale: servicemen were homesick for their families and had not enjoyed home leave in years, they longed »for feminine society,« and they worried that their wives and sweethearts in Britain had forgotten them. Radio was one of the enlisted men's few entertainments, and she found that they listened to it casually in barracks and canteens. Gilbert learned that officers and enlisted men could not hear enough messages and that they even enjoyed hearing messages directed to others. They overwhelmingly favored dance music, crooners over straight singers, and female crooners over male – except for Bing Crosby, who was »in a class by himself.« Their favorite singers were the unrivaled Vera Lynn and Anne Shelton. Gilbert observed, »I came away with the strong impression that apart from a few news bulletins, Ann Shelton and Bing Crosby could more or less fill the entire radio bill between them.«[50]

Within the year, Madden issued a Policy Directive for overseas broadcasts. He warned producers against broadcasting references to leave, suggestions that people at home were »having a very gay time,« alarming air raid stories, »any suggestion that [servicemen's] womenfolk might be getting off with someone else, particularly with troops from Overseas,« and »too much nostalgic sentiment of the ›see-you-home-again-soon‹ type.«[51] *The Star* was the first to publicize the BBC and War Office campaigns against nostalgic and sentimental songs for troops stationed overseas, particularly in North Africa.[52] The response in the press was largely negative: William Hickey of the *Daily Express* called the attack on nostalgia »unduly priggish« and compared it to Nazi banning of »books or art as ›unGerman‹ and ›decadent.‹«[53] The paper's Middle East correspondent, Eric Grey, suggested that the ban on nostalgia was an insult to troops, opining, »They are abroad to do a job and won't go home until it is finished. [...] Naturally their thoughts often turn homeward, but that has never influenced them in their jobs.«[54] After all, letters also reminded the troops of home; the *Daily Express* asked, »Will anyone dare to suggest that the Army Post Office should close down?«[55]

Despite the BBC's fraught relationship with sentimental entertainment, female croonettes, announcers, and producers served an important role in maintaining the morale of their listeners, balancing the sentimental tastes of soldiers with home-front demands for virile music for fighting men. In retrospect, Madden praised their performances at and away from the microphone:

> All the girls took a tremendous amount of trouble to answer those airgraphs and letters, with photos, and they soon became the most popular characters in the respective

> armies, far more so than the variety stars the Forces left behind them, who could not make the same positive contact on the air. This was a war fought with radio: radio for propaganda: radio for morale. These girls did their bit.[56]

Predictably, vociferous individuals at home disagreed when they encountered the fare broadcast to the troops.

The encounter began in February 1944, when the BBC, primarily as a financial measure, replaced the popular (domestic) Forces Programme with the General Forces Programme (GFP), which consisted largely of material broadcast to troops stationed abroad. Among civilians, the BBC paved the way for the new service with a poorly conceived publicity campaign. The BBC let it be known that the notion of a shared GFP to link listeners at home and abroad was inspired by Director-General William Haley's tour of the Middle East, where he had observed servicemen losing touch with their homeland.[57] However, in the same breath that the BBC trumpeted the benefits of the new lifeline, it warned listeners that the GFP was »almost entirely influenced by what your husband or son or brother serving overseas writes to the BBC.«[58] The Corporation's repeated warnings that home listeners' critiques were superfluous proved to be a »psychological mistake« in promoting the new programme, as one radio critic put it – to say nothing of its blithe dismissal of the opinions and tastes of more than half its audience.[59]

Civilians were dubious of the GFP from the start. In a preview of the new programming, much of which had been broadcast overseas for over a year, one syndicated radio reviewer predicted:

> Many new features, strange [...] announcers, and new times for [...] familiar broadcasts will puzzle and intrigue listeners for a time. However, when the novelty wears off, I am afraid many home listeners will wish they could have the old Forces programme back again.[60]

The GFP was soon criticized for including too much boring talk: endless messages programmes, »reiterated news headlines,« and overly chatty female dance band announcers.[61] Some listeners wanted more music of any sort while others complained that in addition to too much talk there was too much »swing and hot music.«[62] A common sentiment was that their common dislike for the GFP would link listeners to their loved ones.[63] Nevertheless, at the end of the first week, Listener Research found that only three percent of civilian listeners opposed the programme, thirty-one percent were favorable, and thirty-five percent reserved judgment.[64]

The two aspects of the new programme that earned the sharpest criticism were women announcers and crooners. Women already played an important role in the home Variety announcing pool, but the GFP featured women newsreaders as well as variety compères. The tradition of criticizing women announcers was sharpened by their greater presence in the GFP. Critics complained that women announcers were thin-voiced, patronizing, and insipid. One listener stated bluntly, »The microphone is his – not hers.«[65] Such comments elicited a speedy defense: radio microphones were modulated improperly for higher voices, women spoke more clearly than men, and general dismissals of women's abilities were motivated by misogyny.[66] Ultimately, the work of women announcers was justified by their popularity with the forces and their power to represent home: a letter to *Christian World* declared, »There is a warmth in their voices, friendliness and anxiety to please, a note of sympathy. [...] I do not doubt that our soldiers, sailors and airmen, wherever they may be, will be cheered and strengthened by the voices of these women from home.«[67]

The most virulent criticisms were reserved for crooners. The debate originated in the 7 March comments of Lord Winterton in the House of Commons, whose words fanned a controversy that lasted over a month. He attacked women crooners: »They remind one of the caterwauling of an inebriated cockatoo... all this wailing about lost babies cannot have a good effect on the morale of the troops.«[68] The voices for and against Winterton's statement were divided fairly evenly. Those who agreed decried the GFP as a programme of demoralizing debauchery. Those who opposed Winterton's words trotted out the usual response that crooning did not harm the morale of troops. »Isn't it a fact that our nation is noted for its splendid morale, and women crooners have been in vogue for some years now?« asked six sergeants.[69] The 1942 ban on crooners should have rendered the problem moot, but it had clearly been ineffective.[70]

More importantly, the troops wanted crooning – and this justified its broadcast: The *Observer* wrote, »Listen to the records demanded of the BBC by the Forces, and slush [...] is shown to be as popular in the camp as ice-cream at the seaside. To cut crooning right out might be a mingling of Attic good taste with Spartan resolution. It would not be democratic.«[71]

The invocation of democracy pointed to the ongoing conundrum of the BBC's wartime broadcasting policy. Would satisfying the desires of listeners serve the needs of a nation at war? What about the forces, their apparent preference for slush, and the potentially dubious effect of sentimentality on morale? The debate over crooning in the GFP debunked more clearly than ever the notion that sentimentality would destroy the morale of the forces, revealing generational and class divisions in musical tastes: the BBC's own logic that troops should determine the content of the GFP undermined its principled 1942 ban on crooners. Rather than muting female crooners as »whining witches« who repelled their boyfriends, Madden led the BBC in recasting female singers, announcers, and producers as unthreatening girl-friends who envoiced the values of home and nation for everyman. Their voices would continue for the next decade to be important in the Light Programme, the postwar successor to the Forces Programme that served the interests of the young and working classes, until the incursions of new musical forms from America recast »the« youth audience.

NOTES

1 Thanks are due to the BBC Written Archives Centre in Caversham (BBC WAC), especially the head archivist Jaqueline Kavanagh, for providing an important service to scholars and permission to reprint BBC copyright material in this article.

2 Letter from Pte. Randall, ET, OFP Section, RAOC, 153 Infantry Brigade, MEF, to Musical Director, BBC, 8 April 1943, BBC WAC Special Collections: Cecil Madden, Wartime Radio Diary, S24/54/19.

3 »Wireless in War,« *The Listener* (7 September 1939): 464.

4 Bill Grieg, »Seven girls keep the Empire's soldiers thinking of home,« *Daily Mirror* (17 February 1942), Madden, Wartime Radio Diary, BBC WAC S24/54/31.

5 See Simon Frith, »The Pleasures of the Hearth: The Making of BBC Light Entertainment,« in *Formations of Pleasure,* ed. Burin Bennett (London: Routledge & Kegan Paul, 1983), 101-23.

6 »Full-Time Programmes for the Forces,« *Radio Times* (16 February 1940): 3.

7 »The Stuff to Give the Troops,« *The Listener* (15 February 1940): 308.

8 »The BBC Has Discovered…,« *Radio Times* (10 March 1939): 11. It did find that age made »a great difference in tastes.«

9 »Full-Time Programmes for the Forces«: 3.

10 Charles Brewer, »What does the R.A.F. listen to?« *Radio Times* (1 March 1940): 5.

11 »Life-Line Between the People at Home and the Men Overseas: BBC Brings Comfort with Its Broadcasts of Messages,« *Clip-Sheet* [Issued by the MoI for the *Exclusive* Use of Editors of Weekly Newspapers], no. 107 (21 January 1942), Madden, Wartime Radio Diary, BBC WAC S24/54/13.

12 Guy Fletcher, »This Week's Radio Miscellany,« *Radio Times* (23 February 1940): 7. Siân Nicholas reports that the practice began at the start of the war when the Home Service broadcast evacuated children giving reassuring messages to their parents. Siân Nicholas, *The Echo of War: Home Front Propaganda and the Wartime BBC, 1939-45* (Manchester: Manchester University Press, 1996), 128.

13 »The Broadcasters,« »Both Sides of the Microphone,« *Radio Times* (20 September 1940): 4.

14 »The Broadcasters,« »Both Sides of the Microphone,« *Radio Times* (8 May 1942): 3.

15 Andrew Crisell, *Understanding Radio* (London: Routledge, 1986), 23.

16 »A Listener Research Report: Listeners' Preferences for Home Service or Forces Programme,« Listener Research Department, 13 January 1942, BBC WAC R9/9/6.

17 Tom Harrisson, »Radio,« *The Observer* (19 July 1942) [BBC WAC Press Cuttings].

18 Ibid.

19 Ibid.

20 »Sloppy,« *Bristol Evening News* (10 July 1942) [BBC WAC Press Cuttings].

21 BTT Lawrence, Lt.-Col., letter to *The Daily Telegraph* (7 March 1942) [BBC WAC Press Cuttings].

22 »A Listener Research Report: Listening Conditions, the Forces Programme Generally and Particular Types of Output in the Forces Programme,« 27 January 1942, Listener Research Department LR/590, BBC WAC R9/9/9; and »Listener Research Bulletin No. 82,« 18 April 1942, Listener Research Department, BBC WAC R9/1/2. The Bulletin included a report on »Listening in the Army Northern Command« by a BBC Public Relations Officer: the survey »covers a very wide variety of units, ranging from an armoured Division to a Military Hospital. In nearly every case general satisfaction with existing programmes was expressed. [...] Much of its value lies in what it does not say. It does not corroborate any complaints about the ›puerility‹ of the Forces Programme which have been expressed in some quarters. It contains no complaints about there being too much light material, nor does it contain any demand for a tougher broadcast diet in the Forces Programme.«

23 »Mike,« »Air Sugar,« *Melody Maker* (4 April 1942): 7.

24 »Radio: For the Princes,« *News Review* (16 April 1942) [BBC WAC Press Cuttings].

25 »BBC Music Policy,« draft by C(P) amended in accordance with suggestions made by DM, DDM and Arthur Bliss, 1 April 1942, BBC WAC R27/73/1.

26 Madden, Wartime Radio Diary, BBC WAC S24/54/15.

27 »Anti Flabby Entertainment in Empire Programmes,« BBC Internal Circulating Memorandum from Mr. Cecil Madden, Criterion, to DEP, 11 March 1942, Madden, Wartime Radio Diary, BBC WAC S24/54/15; and »Extract from Minutes of: Empire Service Meeting,« 11 March 1942, BBC WAC R27/416/1.

28 »Listener Research Bulletin No. 81,« 11 April 1942, Listener Research Department, BBC WAC R9/1/2. Surprisingly, given the programme's iconic status, only thirteen episodes of »Sincerely Yours« aired.

29 Madden, Wartime Radio Diary, BBC WAC S24/54/17.

30 »Anti Flabby Entertainment in Empire Programmes.«

31 Programme listings, *Radio Times* (23 November 1941): 7.

32 Letter from Secretary to Howard Thomas to Clerk of the Council, London County Council, 3 December 1941, BBC WAC R19/683.

33 Madden, Wartime Radio Diary, BBC WAC S24/54/17.

34 »Life-Line Between the People at Home and the Men Overseas: BBC Brings Comfort with Its Broadcasts of Messages.«

35 »Forces Like the Radio ›Girl Friends,‹« *The Evening News* (16 January 1942) [BBC WAC Press Clippings].

36 »Anti Flabby Entertainment in Empire Programmes.«

37 »Sing a Song from London,« *Parade,* Cairo edition (28 February 1942): 16.

38 »Use of Dance Bands in the Overseas Services Notes for C(OS) as Requested,« BBC Internal Circulating Memorandum from Mr. Cecil Madden, Criterion, to Mrs. Spicer, 200 OX ST, 12 November 1943, BBC WAC R27/73/2.

39 Madden, »AEF PROGRAMS do something EXTRA for the Boys,« Madden, Wartime Radio Diary, BBC WAC S24/54/23.

40 Madden wrote, »No-one seemed to mind if a ›Services' Sweetheart‹ was married or had personal attachments. No one really enquired into this and many of the letters to the pin-up girls started frankly by saying that the writer was a married man and wondering if the recipient was married too. However in the main the thread of attraction was the male to the female and the girls got many offers of all sorts.« Madden, »Sweethearts of the Forces« [Memoir], Madden, Wartime Radio Diary, BBC WAC S24/54/17.

41 Madden, »AEF PROGRAMS do something EXTRA for the Boys.«

42 H.S., »Miscellaneous and Dance,« *The Gramophone* (February 1943): 131.

43 Letter from Captain WM Mair, Chopel Sultania Infantry, MEF, to Anne Shelton, 13 August 1941, Madden, Wartime Radio Diary, BBC WAC S24/54/9.

44 Capt. CF Milner, letter to *Radio Newsreel,* (13 September [1942]) [BBC WAC Clippings Files].

45 »Lili Marlene,« BBC Internal Circulating Memorandum from Mr. Cecil Madden, Criterion, to DEP, 8 August 1942, BBC WAC R27/178.

46 »Anne Shelton: *Introducing Anne* Series,« BBC Internal Circulating Memorandum from Mr. Cecil Madden to DEP, 7 October 1942, BBC WAC R19/1153/1.

47 »Anne Shelton,« BBC Internal Circulating Memorandum from Mr. Alick Hayes to DEP, 7 October 1942, BBC WAC R19/1153/1.

48 Mavis Nicholson, *What Did You Do in the War, Mummy?: Women in World War II* (London: Chatto & Windus, 1995), 12. Goebbels, incidentally, considered the song to be morale sapping, describing it as having a »cadaverous smell,« and attempted unsuccessfully to ban both song and Andersen when the Gestapo caught the singer attempting to escape Germany in the summer of 1942. Horst J.P. Bergmeier and Ranier E. Lotz, *Hitler's*

Airwaves: The Inside Story of Nazi Radio Broadcasting and Propaganda Swing (New Haven: Yale University Press, 1997), 188.

49 Joan Gilbert, »Report on Visit to Gibraltar,« February 1943, Madden, Wartime Radio Diary, BBC WAC S24/54/19.

50 Gilbert, »Report on Visit to Gibraltar.«

51 Producers also needed to avoid racist language that might offend allies. Madden, »Policy Directive for Variety Material Broadcast Overseas,« [c. 6 October 1943] Madden, Wartime Radio Diary, BBC WAC S24/54/19.

52 [Untitled], *The Star* (9 April 1943) [BBC WAC Press Cuttings].

53 William Hickey, »Should Slush Be Shushed?« *Daily Express* (5 April 1943) [BBC WAC Press Cuttings].

54 Eric Grey, »›Homesick‹ Songs Do Not Worry the Middle East Army,« *Daily Express* (8 April 1943) [BBC WAC Press Cuttings].

55 [Untitled], *Daily Express* (9 April 1943) [BBC WAC Press Cuttings].

56 Madden, »Sweethearts of the Forces« [Memoir].

57 Rebecca West, »The BBC Puts Over a Bad Egg,« *Picture Post* (18 March 1944): 19 [BBC WAC Press Clippings].

58 George Stagg, »These Men Make the BBC Sit Up and Take Notice,« *Daily Express* (5 February 1944) [BBC WAC Press Cuttings].

59 »Radio Critics,« *Evening Chronicle* (8 March 1944) [BBC WAC Press Clippings].

60 Ronald Strode, »You Will Hear Stuff They Give the Troops,« *Sunday Sun* (27 February 1944) [BBC WAC Press Clippings].

61 Jonah Barrington, »BBC Should Drop the ›Forces,‹« *Sunday Chronicle & Sunday Referee* (2 April 1944) [BBC WAC Press Clippings]; and Collie Knox, »Cut the Cackle and Get to the Forces,« *The Star* (17 March 1944) [BBC WAC Press Clippings].

62 »Regular Listener,« letter to *Ashton-under-Lyne Reporter* (17 March 1944) [BBC WAC Press Clippings]; and »For Mr. Foot,« *The Star* (7 March 1944) [BBC WAC Press Clippings].

63 West, »The BBC Puts Over a Bad Egg.«

64 17% were »not interested« and 13% were unaware of the change. »Last Word Still the Forces'« *Sunday Express* (12 March 1944) [BBC WAC Press Clippings].

65 »A Listener,« »I Hate Women Broadcasters,« *Daily Mail* (15 February 1944) [BBC WAC Press Clippings].

66 »Women on the Air: Our Readers' Verdict,« *Daily Mail* (25 February 1944) [BBC WAC Press Clippings].

67 »This is London!« *Christian World* (16 March 1944) [BBC WAC Press Clippings].

68 »Women Crooners Say ›We Don't,‹« *Evening News* (8 March 1944) [BBC WAC Press Clippings].

69 Six Sergeants in Orkney, letter to *Aberdeen Press and Journal* (20 March 1944) [BBC WAC Press Clippings].

70 »Crooning for All,« *Nottingham Guardian* (8 March 1944) [BBC WAC Press Clippings].

71 »Soft and Low,« *The Observer* (12 March 1944) [BBC WAC Press Clippings].

REFERENCES

BBC Written Archives Centre, Caversham: Radio Files and Special Collections:

R9/1/2: Audience Research/Bulletins, Sound/69-121 (1942).

R9/9/6, 9: Audience Research /Special Reports, Sound and General, files 6 and 9 (1942, 1945).

R19/683: Entertainment/Vera Lynn Programmes (1941-1953).

R19/1153/1: Entertainment/Anne Shelton Programmes (1942-44).

R27/73/1-2: Music General/Dance Music Policy, file 1 (1941-42)-file 2 (1943-44).

R27/178: Music General/Lili Marlene (1942-47).

R27/416/1: Music General/Overseas Service, file 1 (1936-42).

S24/54/9, 13, 15, 17, 19, 23, 31: Special collections/Madden/Wartime Radio Diary.

Contemporary Press Sources:

Barrington, Jonah. »BBC Should Drop the ›Forces.‹« *Sunday Chronicle & Sunday Referee* (2 April 1944). [BBC WAC Press Clippings.]

»The BBC Has Discovered…« *Radio Times* (10 March 1939): 11.

Brewer, Charles. »What does the R.A.F. listen to?« *Radio Times* (1 March 1940): 5.

»The Broadcasters.« »Both Sides of the Microphone.« *Radio Times* (20 September 1940): 4.

»The Broadcasters.« »Both Sides of the Microphone.« *Radio Times* (8 May 1942): 3.

»Crooning for All.« *Nottingham Guardian* (8 March 1944). [BBC WAC Press Clippings.]

Fletcher, Guy. »This Week's Radio Miscellany.« *Radio Times* (23 February 1940): 7.

»For Mr. Foot.« *The Star* (7 March 1944). [BBC WAC Press Clippings.]

»Forces Like the Radio ›Girl Friends.‹« *The Evening News* (16 January 1942). [BBC WAC Press Clippings.]

»Full-Time Programmes for the Forces.« *Radio Times* (16 February 1940): 3.

Grey, Eric. »›Homesick‹ Songs Do Not Worry the Middle East Army.« *Daily Express* (8 April 1943). [BBC WAC Press Cuttings.]

HS. »Miscellaneous and Dance.« *The Gramophone* (February 1943): 131.

Harrisson, Tom. »Radio.« *The Observer* (19 July 1942). [BBC WAC Press Cuttings.]

Hickey, William. »Should Slush Be Shushed?« *Daily Express* (5 April 1943). [BBC WAC Press Cuttings.]

Knox, Collie. »Cut the Cackle and Get to the Forces.« *The Star* (17 March 1944). [BBC WAC Press Clippings.]

»Last Word Still the Forces.« *Sunday Express* (12 March 1944). [BBC WAC Press Clippings.]

Lawrence, Lt.-Col. BTT. Letter to *The Daily Telegraph* (7 March 1942). [BBC WAC Press Cuttings.]

»A Listener.« »I Hate Women Broadcasters.« *Daily Mail* (15 February 1944). [BBC WAC Press Clippings.]

»Mike.« »Air Sugar.« *Melody Maker* (4 April 1942): 7.

Milner, Capt. CF. Letter to *Radio Newsreel* (13 September [1942]). [BBC WAC Clippings Files.]

Programme listings. *Radio Times* (23 November 1941).

»Radio Critics.« *Evening Chronicle* (8 March 1944). [BBC WAC Press Clippings.]

»Radio: For the Princes.« *News Review* (16 April 1942). [BBC WAC Press Cuttings.]

»Regular Listener.« Letter to *Ashton-under-Lyne Reporter* (17 March 1944). [BBC WAC Press Clippings.]

»Sing a Song from London.« *Parade,* Cairo edition (28 February 1942): 16.

Six Sergeants in Orkney. Letter to *Aberdeen Press and Journal* (20 March 1944). [BBC WAC Press Clippings.]

»Sloppy.« *Bristol Evening News* (10 July 1942). [BBC WAC Press Cuttings.]

»Soft and Low.« *The Observer* (12 March 1944). [BBC WAC Press Clippings.]

Stagg, George. »These Men Make the BBC Sit Up and Take Notice.« *Daily Express* (5 February 1944). [BBC WAC Press Cuttings.]

Strode, Ronald. »You Will Hear Stuff They Give the Troops.« *Sunday Sun* (27 February 1944). [BBC WAC Press Clippings.]

»The Stuff to Give the Troops.« *The Listener* (15 February 1940): 308.

»This is London!« *Christian World* (16 March 1944). [BBC WAC Press Clippings.]

[Untitled.] *Daily Express* (9 April 1943). [BBC WAC Press Cuttings.]

[Untitled.] *The Star* (9 April 1943). [BBC WAC Press Cuttings.]

West, Rebecca. »The BBC Puts Over a Bad Egg.« *Picture Post* (18 March 1944): 19. [BBC WAC Press Clippings.]

»Wireless in War.« *The Listener* (7 September 1939): 464.

»Women Crooners Say ›We Don't.‹« *Evening News* (8 March 1944). [BBC WAC Press Clippings.]

»Women on the Air: Our Readers' Verdict.« *Daily Mail* (25 February 1944). [BBC WAC Press Clippings.]

Secondary Books and Articles:

Bergmeier, Horst J.P. and Ranier E. Lotz. *Hitler's Airwaves: The Inside Story of Nazi Radio Broadcasting and Propaganda Swing.* New Haven: Yale University Press, 1997.

Crisell, Andrew. *Understanding Radio.* London: Routledge, 1986.

Frith, Simon. »The Pleasures of the Hearth: The Making of BBC Light Entertainment.« In *Formations of Pleasure.* Edited by Burin Bennett. London: Routledge & Kegan Paul, 1983. 101-23.

Nicholas, Siân. *The Echo of War: Home Front Propaganda and the Wartime BBC, 1939-45.* Manchester: Manchester University Press, 1996.

Nicholson, Mavis. *What Did You Do in the War, Mummy?: Women in World War II.* London: Chatto & Windus, 1995.

Selling the War in Iraq: Television News Music and the Shaping of American Public Opinion[1]

JAMES DEAVILLE

If we were to view any of the »stingers« at the opening of network news coverage from the »Shock and Awe« campaign of the War on Iraq, we would have to find the images striking. We might even perversely call them »spectacular.« But the effect of these news production elements is not purely visual: a whole sonic realm works with the images to produce their impact. Musicians may have observed the role of music in creating affect in the stingers, but how many members of the American public-at-large, the consumers of such media representations, realize that the music is selling us a certain take on the news? In the context of television, this political application of music, the »ultimate hidden persuader« to quote Nicholas Cook,[2] is all the more powerful not only by virtue of the size of its audience, but also because it seems natural.

Indeed, a multi-million dollar industry is devoted to providing news music for the broadcast media.[3] Normally, local radio and television stations and even national networks do not possess the resources to have a composer on staff and, thus, they turn to production companies that sell licensed packages or libraries of musical material, called elements, for insertion at appropriate moments within the broadcast. The most popular of such production elements in television news include:[4] 1) Opens and closes, also known as titles or themes, identify the newscast and usually provide musical material for the rest of the broadcast – one can find them in newscasts around the world.[5] Through fanfare-like, wide-ranging, quick, syncopated themes in brass and strings over a driving beat, they create the impression of a dynamic newsroom, while at the same time establishing an authority in the eyes and ears of the audio-viewer.[6] 2) Stingers or teases are very brief musical and visual markers for an important news item; occurring just before specific coverage begins, they serve the dual purpose of notifying the public about the importance of the reports to follow and – subliminally – helping to create an attitude within the viewership that will not only accept but also adopt the broadcast's position. These are the most original and free production elements of the newscast.[7] 3) Beds, often called »promo beds,« provide framing and background music for extensions of opens or longer previews of upcoming items, under voices and images. Beds at the beginning of newscasts often take the opening theme and loop several of its measures in the middle, so that a highly variable length is possible. 4) Bumpers are the musical lead-ins to and lead-outs from commercials, which like stingers create a mood for the news items to which they refer. Unlike opens/closes and beds, bumpers within one newscast may widely differ, depending on the topic of the referenced news item.[8]

Thus standard signifying practices in this music industry have created designations that describe either the placement of the music (e.g. »opens«) or its affect (»teases«), which foregrounds the role and importance of music as »hidden persuader.« As Larry the O. from LucasArts observes, listening to this music »is not a conscious thing; it's subliminal… The whole idea behind sound design… is that the [viewer…] buys into the whole thing.« Lawrence Grossman, former president of NBC News, put it this way with regard to the news music of CNN and Fox: »[It] tells you what to think.«[9] The industry relies upon musical signifying practices that are readily understood, and yet seem natural to the recipient. Such practices are nothing new to music, whether embodied in theories explaining the musical »doctrine of affections« of the Baroque Era or the affectively organized anthologies of music for silent films from the beginning of the twentieth century.[10]

Production companies of today fully subscribe to music as the bearer of messages, and if there is any question, we can just look to their own, often disturbing rhetoric. Thus the British firm Mokal Music advertises its broadcast package called Shuffle with the words »mean and pushy, strong & persuasive. These themes cover a lot of beefy styles, all up-tempo and up attitude.«[11] And the track entitled »This Evening« is described as »persuasive theme featuring solo sax, horns, and guitar. Perfect for newscasts…«[12]

Shelly Palmer Productions, Inc. of New York City sells to radio and television stations news music packages that are guaranteed to have an effect upon listeners and viewers.[13] In the significantly named »Brave New World,« the productions elements are alternately described as »majestic,« »hard-hitting,« »jazz-oriented« and »timeless,« which suggest a full range of musical expression and moods.[14] Designations like »Crime« and »Coping« make the signification of specific 4½ second cuts clear,[15] musically corresponding to our expectations from the descriptive titles.

These examples prove that the networks have come to recognize for their news broadcasts what television advertising agencies have long known: music is unsurpassed in its ability to tap into the personal narrative of the viewer and to suture her or him into the diegesis. As Claudia Gorbman observes in her landmark book *Unheard Melodies*, »music may act as a ›suturing‹ device…, lessening awareness of the technological nature of film discourse. Music gives a ›for-me-ness‹ to the soundtrack… [And] music increases the spectator's susceptibility to suggestion.«[16] Put in slightly different terms, if music is an important component in the construction of personal narrative, and the newscast is geared to convince the viewer that its purpose coincides with his or her narrative (»following the stories that are important to YOU« or »keeping an eye on YOUR city«), then this can create a commonality so that the audio-viewer regards the music chosen for a particular production element as natural and his or her own choice.

It is not really any jump at all to the use of music to sell not only one particular network's coverage of a war, but also an attitude about that war, whereby American media empires seem to be marching in lockstep with the government. As media personality Danny Schechter writes, »it started with the Gulf War – the packaging of news, the graphics, the music, the classification of stories… If you can get [a sedated public] hooked…, you have a winner in terms of building audience.«[17] War may »unleash devastation and death on people, [but it also] delivers ratings and brings life to television. War is often the ›big story‹ (when sex isn't) and a defining moment for many journalists…« And for the American public, it is one's patriotic duty to support a just war against terrorists who have heinously attacked the homeland.

Indeed, in the US media since 9/11, we can observe almost a continuous beating of war drums and sounding of martial trumpets, after the initial sounds of grief and mourning gave way in late September to the bellicose tone in broadcast media coverage of the War on Terrorism. As noted media researcher Shoma Munshi observes, »from the days and weeks following 9/11, till the United States attacked Afghanistan in October 2001, and most recently Iraq in March 2003, US broadcast media have managed to maintain a sustained level of patriotism fanning the public mood to keep the United States safe by whatever means possible.«[18] Already shortly after 9/11, NPR asked its composer Jeffrey Freymann-Weyr for special coverage music, »either for another crisis in the States, or, as it happened, an attack on Afghanistan.«[19] The music he composed for the events of October 2001 found use again in March of 2003, albeit with »subtle changes« dictated by the network. We shall return to this specific music, but we should note that even a politically liberal news provider like NPR, which was no uncritical supporter of the war on terrorism, recognized the value of creating a »war music« that would engage the public.

Thus if the evening of 19 March 2003 brought the unannounced, yet not unexpected, onset of the War on Iraq, the networks were well prepared. Five days before the war began, Fox News had its music package entitled »Liberation Iraq Music« ready to go.[20] Bob Israel of Score Productions, the source of the ABC Nightly News theme in 1978, used the same thematic material to make a soundtrack for the ABC war news in advance of March 19th. When queried about preparing a soundtrack for a war that was not yet certain, he explained that it »sounds a little crass, but that's what you have to do in this business to be prepared.«[21] Although the dates are unclear, CBS had commissioned Peter Fish, composer of the Evening News theme, and NBC had done the same from Michael Karp, creator of the *Dateline NBC* sound, to write their War In Iraq music before the conflict began. NBC told Karp, »it looks like there's going to be a war and we could use a good theme.«[22] And in anticipation of finding a buyer from some station or network, music production company TM Century had already created the militaristic music package »Juggernaut« in December 2002, under the assumption – according to Nicholas Engstrom in the *Columbia Journalism*

Review – that »within six months there would be either a war or another major terrorist attack on American soil.«[23]

What were network executives looking for when they commissioned this music? *The Philadelphia Inquirer*'s music critic Peter Dobrin summarized the desired effects: »Like quick-firing subliminal messengers, special music is telling TV viewers and radio listeners what to think and how to feel about the war on Iraq.«[24] Thus Richard O'Brien, Fox News's Creative Director, wanted something »with a marching feeling... We wanted the music to say, ›Something big is coming this way.‹« After previewing the package, O'Brien directed the unnamed composer to »put in more tom-tom drums because they had more urgency. I wanted it to sound like, I don't want to say war drums, but...«[25] When the director of the CBS Evening News, Eric Shapiro, asked Peter Fish to write music that »tries to take into account as many situations as possible« and »conveys some idea of mood,« the composer responded with war music that – in his words – conveyed a »climate of fear. To me, this is a real live war, and we should be both awed and simultaneously scared... It is just Techno-Ali-vs.-Frasier-IV-we're-going-to-knock-the-crap-out-of-them, testosterone-driven, big-punch music.«[26] Personally, Fish attests to being a strong opponent to the war, »nevertheless, I have my job to do... I believe the general mood of this country at this point in time is ›we're going to go kick some Arab ass.‹«[27]

Fish's theme for CBS was first aired on 19 March 2003 at 6:33 p.m. EST, the onset of the war. It is awful, also by the standards of television news music. Baer calls it artificial – »full of synthesizer drum beats, the sound of helicopters, and superficial thematic material; the rhythms are repetitive.«[28] One could also complain about its almost inaudible thematic material and cheap sound effects. This is not the type of war theme that is likely to evoke either a subliminal or an aggressive response.

Upon comparing Fish's package with the music written by Karp for NBC, we discover how significantly the war theme and resulting package differed between networks. NBC wanted »that kind of classy signature sound that NBC seems to have over many of the other networks.«[29] Karp managed to capture the symphonic sound of John Williams's signature music for »The Mission,« while enhancing the material's inherent martial qualities through bells and strings.

Quickly moving violin parts give the music a sense of urgency, given intensification by adding another string line in thirds above the original. This ennobling music can be all the more effective in terms of shaping attitude because it is not in your face like the CBS theme or Fox's war music.

As a brash upstart, Fox News does not have the same image to uphold as do the established networks, so it did not have to employ anything remotely dignified in its war coverage. The aforementioned music with the drums of war has been described by Adam Baer in *Slate* as a »tanked-up shock-and-awe campaign of gung-ho missile ›whooshes‹ and high-pitched electro-shrieks. Inspired by rock 'n 'roll... it sounds youthful, impatient, and reactionary.«[30] Engstrom writes that the music »could be Metallica rehearsing Wagner...«[31] Further description is not really necessary for the highly synthesized, pounding, even violent music that has been closely edited to the video images. There is nothing subtle about this score, which John von Rhein of the *Chicago Tribune* describes as »trumpeting an unequivocal, scorched-earth patriotism: Let's feel good about America and beat Saddam!«[32]

If these networks, whatever their orientation, seemed to march musically in lockstep with the Iraq war politics of the Bush administration, one network resisted the pressure to join the bandwagon (albeit in its war music for radio). NPR's director for *All Things Considered*, Bob Boilen, commissioned Freymann-Weyr to write »more compassionate music, more thoughtful without being sad, because no matter how you feel about this conflict, I think people feel compassion for the soldiers and the innocents.«[33] As Freymann-Weyr echoed, »the challenge was that it needed to be serious but not gloomy, not overly militaristic or flag-waving.«[34] The result is a subdued, static music that, other than its few nods to tradition, flies in the face of every other war music from March 2003. With its reflective, uncertain tone, the music certainly differs from that of the other networks, which Freymann-Weyr explains in the only interview that details a composer's approach to writing music for the War on Iraq: »One thing I tried to do with the harmony was to introduce the ambiguity that (I hoped) would keep it from going too far in either direction – overly traditionally patriotic or overly morose. Even though the piece is centered around the key of C, I purposely avoided the one note that would make it either major or minor – all of the C chords, instead of E or E-flat, have a D and F natural instead.«[35] Still, he felt he needed to connect with broader principles of signification for war music: »Although trumpet, timpani and military snare drums are a bit of a cliché, it didn't feel right *not* to use them, given the history of music in times of war.« The result is a plaintive theme music that signifies the human tragedy of war. Of course, NPR's radio coverage had the advantage of not being accompanied by graphics and video images that might influence the emotional tone adopted by the music.

The reality of the first days of the war confounded network expectations that the early going would provide spectacular visuals from the shock-and-awe campaign. According to Engstrom, music played a greater role at the beginning, because »the first couple of days were less visually dramatic than anticipated.«[36] In comparison with similar footage from the Persian Gulf War and the bombing of Baghdad, one can concur that the absence of a sustained bombing campaign in Baghdad made for visually uninteresting television screens in the homes of America, as citizens hoping for spectacular surgical strikes audio-viewed silent, dark skies over Baghdad or heard bomb blasts without the accompanying fireworks. In fact, on 19 March 2003 at 8:41 p.m., Dan Rather of CBS was clearly caught off-guard when, as he was reporting against a live video of the nighttime Baghdad skies, deafening explosions sounded without any visual markers. The pre-eminence of sound over sight in creating believability here might lead us to regard this media moment as »auricular« rather than »spectacular.«

However, Engstrom does not account for the difference between live coverage, which did not offer much to the senses, and the network packaging of the news from the front, which from the start used music and imagery to enhance the message of a justified conflict that »we« can win. We have already seen how networks put together music, graphics and video footage in stingers and bumpers at 6:30 and other times of breaking news to brand their particular take on the war.[37] The one element added to this television news media mix that exponentially increased the danger for the audio-viewer were the actual sounds of war. When sound editors mixed the sounds of air-raid sirens, weapons fire, and jet take-offs with a music track in the opening bed of the newscast, the result naturalized the violence or – in the words of music critic von Rhein – »sanitized war's darkest realities.«[38] As we can observe in any of the stingers or beds from the War on Iraq, sounds and music become one in the soundtrack for war, on the one hand bringing the danger, excitement, »liveness« of the armed conflict into our living rooms,[39] on the other presenting the invasion as a media event, a glamorized, »staged spectacle.«[40] Moreover, as a matter of principle, editors enhance the sounds of war to add to the spectacle (or auricule). Indeed, sound effects can stand in for music to create dramatic moments in war coverage, which the CBS example illustrated. If one of the purposes of the aural component of television is to suture the audio-viewer into the diegesis, how much more do the digitally enhanced sounds of screaming jets, shrilling sirens and booming explosions make us feel that we are there, that it is our war, that we must support it, that we must win it?

But when the bombs are no longer our own and the reason for the conflict is not clear any more, when public opinion has turned against the conflict, the music and sounds of war disappear. Or do they? For a newscast segment called »Fallen Heroes« that has aired daily since early May of 2004, Fish turned his CBS signature theme into a plaint for the featured military deceased, through the use of solo winds (especially clarinet and oboe), soft dynamics and slow tempo. With this music, we return to a world before the War on Terror, seeking solace for acts of violence against Americans. Unlike the Persian Gulf War, there has been no closure to the War in Iraq, no winner or loser, not even a clear sense of good guys and bad guys, at least since the prisoner-abuse scandal at Abou Ghraib.

Where is the true threat to freedom, in the non-existent weapons of mass destruction or in the very real, yet hidden, weapons of mass deception?

NOTES

1 My thanks extend first and foremost to the staff of the Vanderbilt Television News Archive, who enabled me to conduct on-site research into this topic during two visits in 2003 and 2004. Without their holdings of daily network newscasts and hourly coverage of breaking news stories like the War on Iraq, this type of investigative research would not be possible. Valuable comments have been contributed by Anahid Kassabian (Fordham University), Robynn Stilwell (Georgetown University) and Michael Saffle (Virginia Tech), whom I thank for their generous assistance.

2 Nicholas Cook, *Music: A Very Short Introduction*, 2nd ed. (Oxford: Oxford University Press, 2000): 122.

3 Unfortunately, this industry has not received any scholarly attention to this point in time, despite its significance in shaping musical discourse of the late twentieth century. The commercial basis and disappearing nature of music for the news have conspired to render it invisible to musical scholarship.

4 These »definitions« are better termed descriptions, since there exists no authoritative glossary of the industry. The descriptions are based upon observations of a variety of websites for television production element providers, even though terms may well vary from one company to another.

5 Interestingly, opens from newscasts of other countries (visual and sonic elements) are often presented in North American reporting about the contents of those foreign newscasts. This practice must establish a certain authenticity for the North American representation of the »exotic other.« That music integrally contributes to the establishment of identity here bespeaks its effectiveness as »hidden persuader.«

6 Michel Chion coined the term »audio-viewer« for the consumer of film and television, in order to foreground the (or at least bring into balance) the sonic aspect of the media, which disappears through the traditional designations of »viewing« or »watching.« See his *Audio-Vision: Sound on Screeen*, trans. Claudia Gorbman (New York: Columbia University Press, 1995).

7 Probably the best-known stinger is associated with CNN's coverage of the Persian Gulf War: a tympani roll that decrescendos, against a visual of a tank that quickly comes into focus. The very simplicity of the music creates a believable enemy, whom we vanquish with an authentic cadence.

8 Bumpers in television news music have the same effect as the act-ins and act-outs of network serials, since they help mask transitions into and out of commercial breaks. They do not figure in continuous, commercial-free newscasting after a crisis, such as the attacks of 9/11, although networks might use bumpers before and after advertising for their own coverage of the crisis.

9 Nicholas Engstrom, »The Soundtrack for War,« *Columbia Journalism Review* 42.1 (March 2003): 45.

10 Among the most noted such collections is Erno Rapee's *Encyclopedia of Music for Pictures* (New York: Belwyn, 1925), which is a compendium of hundreds of suggestions for musical cues, drawn from the classical and light-classical repertoires and arranged by mood and location.

11 »Production Music Demo Catalog,« britishaudio.com, p. 2, at <http://www.britishaudio.com/Downloads/2002_BA_Catalog.pdf>. Accessed on 11 June 2005.

12 <http://www.britishaudio.com/Production_Music_CD/MML-102Shuffle.htm>. Accessed on 11 June 2005.

13 Under their »Custom News Music & Image Campaigns« web page, Shelly Palmer Productions writes the following: »You know our most popular themes… we do more news and station image music than just about anyone. Our syndicated news themes are timeless equity-builders. We contemporize and expand our packages on a regular basis so they never go out of style… We're the news music experts… just listen!« <http:www.shellypalmer.com/Pages/station3.com>. Accessed on 11 June 2005.

14 <http://www.shellypalmer.com/pages/bnw.htm>. Accessed on 11 June 2005.

15 Shelly Palmer Productions offers, for example, three different themes described as »Crime,« each with eight varying orchestrations according to the time of day and type of newscast. They also can provide the enterprising station with two »Coping« themes, this time with seven variants for each.

16 Claudia Gorbman, *Unheard Melodies: Narrative Film Music* (Bloomington, IN: Indiana University Press, 1987): 5.

17 Matthew A. Baum, »Sex, Lies, and War: How Soft News Brings Foreign Policy to the Inattentive Public,« *American Political Science Review* 92.1 (March 2002): 105.

18 Shoma Munshi, »Television in the United States from 9/11 and the US's Continuing ›War on Terror‹: Single Theme, Multiple Media Issues,« *Media, War, and Terrorism: Responses*

from the Middle East and Asia, ed. Peter van der Meer and Shoma Munshi (New York: Routledge, 2004): 50.

19 Jeffrey Freymann-Weyr, »NPR's Special Coverage Theme Music.« <http://www.npr.org/news/specials/iraq2003/theme.html>. Accessed on 11 June 2005.

20 Engstrom 45.

21 Ibid. 46.

22 Carter Burwell, »Music at 6: Scoring the News, Then and Now,« *Esopus* 1 (Fall, 2003): 37.

23 Engstrom 45.

24 Peter Dobrin, »Media's War Message Carries a Message,« *The Philadelphia Inquirer*, Sunday, 30 March 2003.

25 Engstrom 47.

26 Burwell 37.

27 Ibid. 38.

28 Adam Baer, »The Sounds of War – Rating the News Networks' Theme Music,« *Slate*, Thursday, April 17, 2003. <http://slate.msn.com/id/2081608>. Accessed on 11 June 2005.

29 Burwell 37.

30 Baer, »The Sounds of War.«

31 Engstrom 45.

32 John von Rhein, »U.S.A. Networks' Theme Music Sanitizes War's Darkest Realities,« *Chicago Tribune*, Saturday, 6 April 2003.

33 Dobrin, »Media's War Music Carries a Message.«

34 Freymann-Weyr, »NPR's Special Coverage Theme Music.«

35 Ibid.

36 Engstrom 46.

37 By March 21, every network other than CNN had abandoned its continuous coverage of the war.

38 von Rhein, »U.S.A. Networks' Theme Music Sanitizes War's Darkest Realities.«

39 See Philip Auslander, *Liveness: Performance in a Mediatized Culture* (London: Routledge, 1999): 10-22 for an extended discussion of television as a live medium.

40 Danny Schechter, *Embedded: Weapons of Mass Deception: How the Media Failed to Cover the War on Iraq* (Georgetown, ON: COLDTYPE.NET, 2003): 19.

T-Shirts: Local and Global Connotations of a Deceptively Simple Piece of Textile

PETER STACHEL

The story of a life can be told in t-shirts. (Brunel 13)[1]

The most ingenious piece of cloth since the fig leaf. (Commercial for Brunel's book)

T-shirts might be the most common item of clothing one can think of. They are simple, just a basic cotton shirt that covers the body and usually also the shoulders, and they are worn nearly all over the world by people of nearly all cultural backgrounds and habits, all social classes, ages, sexes and sexual orientations. In a way they seem to be the most egalitarian of clothing, but as surfaces for iconic and/or verbal messages, they can be used to express the individual or group identity of the person who wears them, and very often they are linked with places and/or companies. On the other hand, the elements to express individual or local identity are quite similar all over the world. While the message might be a local or personal one, the way to express it is more or less formalized so that t-shirts can be seen as a typical cultural element of »glocalization,« the interpenetration of the global and the local, meaning that both must be attended to at the same time.

While t-shirts were originally elite male underwear, nowadays they are for everyone and their dog, literally. Worldwide about two billion t-shirts are now sold a year. As the magazine *Elle* wrote in the 1970s, t-shirts will never fall out of fashion because they never were fashion in the first place. But exactly that fact – that t-shirts are basics – makes them relevant for research. They belong to those objects that we all own, see daily and know, mostly without thinking about them. However, as the conservative Swiss art-historian Beat Wyss wrote in his 1997 book *Die Welt als T-shirt* (*The World as T-Shirt*), the t-shirt is the typical emblem of our time, just as the wig was in the eighteenth century. This contribution will establish that wearing t-shirts is a historically specific cultural habit that tells us something about the cultural situation we are living in.

The typical t-shirt, called the basic, has the form of a »T« – the origin, of course, of its name – and is made of 100% cotton. The classic basic is white without any color or ornamentation; however, many different colors and styles of t-shirts have emerged: the so-called »tank top« without sleeves or, usually for women, with spagetti straps; shirts with long sleeves, which at one point shifted into heavier sweatshirts; shirts embossed with different kinds of images, prints, sequins and studs, sometimes also in relief; net shirts; tie-dying, which was popular in the late 1960s and early 1970s, and so on. The basic t-shirt, however, is what the older German-speaking ethnology would have called an *Ur-Tracht*: a piece of textile, of most simple, even primitive form. The basic t-shirt is light and thin, absorbent, easy to clean and can be rolled to a small ball without being damaged. It is not only easy to wear and care for; it is also easy to produce and to decorate. Not only a piece of cloth, it is a surface which invites images and messages. In short, it is a medium. It is easy, and not even expensive, to produce one's own personal t-shirt by placing an image or a slogan on a basic shirt. This makes it possible to react very quickly and produce very current images and messages. In our world, formed as it is by the global economy and mass-consumption, consumers try to act and express themselves as individuals. Even when they use mass-products, they try to accommodate them as expressions of their own identity. For that purpose, t-shirts are ideal. In the words of Charlotte Brunel:

> In times, when social connectedness becomes weaker and the role of the individual citizen in political debates becomes less important, the t-shirt is more than the individual medium of expression of one single individual. The person who is a product of the consuming society sees him/herself as a rebel. […] As an integrated part of our society, dominated as it is by the regulations of the market, the t-shirt coins fashions and trends and becomes part of remembering in a world that changes faster and faster. Automati-

> cally it becomes a sort of anchor. Yesterday's promotion-shirt reminds us of a past that quickly becomes forgotten. As a souvenir, the t-shirt becomes loaded with feelings. One keeps it like a relic, uses it as pajamas or – throws it away quickly. In any case – t-shirts communicate. (Brunel 12, 384)

T-shirts can be, and are, used as a medium of communication. They can be used to attempt to establish contact with people – for example, for sexual relations, as many sayings on t-shirts indicate. They are used as advertisements and to express political statements. They can also remind us of past periods of our personal life, a holiday long ago, a time we spent at a university in another country, or even a conference we took part in. In the words of Charlotte Brunel: »The story of a life can be told in t-shirts« (13).

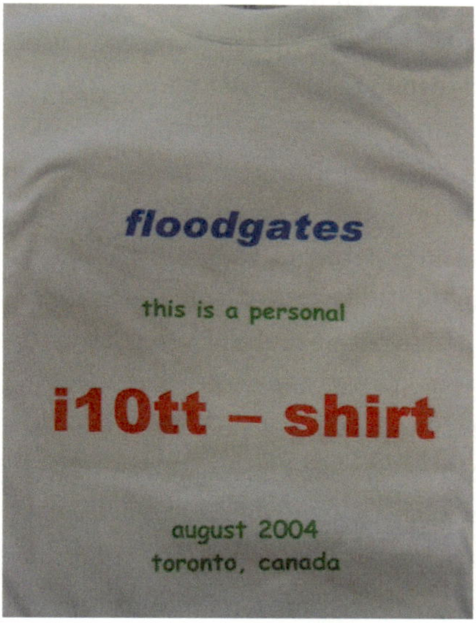

Illustration 1: Personal t-shirt from the »Floodgates« conference, Ryerson University, Toronto, Canada, August 2004.

Most t-shirts are not produced in the same place they are sold, and this also affects shirts linked to places by the images printed on them, imbuing them with social capital. Most t-shirts are currently produced in Asian or Latin American countries, as are most which carry other information on the label. It is worth noting that a basic, raw shirt – without print and ornamentation, directly sold from the producer – costs no more than about 30 cents.

The increasing popularity and ubiquity of the t-shirt is a result of changes in clothing that reflect both changes in social images of the body and socio-cultural changes in modern societies, marked by the shifting of social borders, of gender boundaries,[2] of the boundaries between private and public life and of a radical change in public taste.[3] When t-shirts were simply male underwear[4] and cheap cotton-shirts not yet produced industrially, underwear was a luxury that only the higher classes could afford. The industrial processing of cotton made this sort of underwear more affordable. Besides being worn as underwear, the t-shirt was also worn as working clothes by the lower classes – wearing these shirts in public was coded typically proletarian. Until 1945 in the United States, the market for underwear t-shirts was dominated by two companies that are still well known: »Hanes« and »Fruit of the Loom.« The latter has been using its well-known logo since 1875. The shirts, white basics, were commonly sold in packages of three.

Around the beginning of the twentieth century, shirts like the t-shirt began to be worn for sports as well, becoming in the process a social phenomenon not only of the upper classes but of society at large. From the 1930s on, professional sport teams used t-shirts, for example, in soccer and in baseball. At that time in the US, baseball t-shirts with a number and the name of the player on the backs were used for the first time – nowadays a classical image. Shirts of that kind were also connected with marine images, as sailors (e.g. US marines) wore such shirts to work in. In the 1930s, these shirts were called »Gob-Shirts,« which is linked with the image of a marine in the same way the contemporary German, especially Austrian-German, word *Ruderleiberl* (rowing shirt) was.

The global success story of the t-shirt started during the Second World War, when US troops in the Pacific began to wear these shirts instead of uniforms, which were very unpractical in that climate. From 1942 on, the t-shirt – first in white, later in khaki – was officially part of the uniform of US soldiers, and for that reason it needed an official name, which was provided by the US Army: »T-style-shirt.« In the catalogues of Sears & Roebuck at that time, the shirts were called »Army style t-shirts,« and that is where the name that is now so common comes from. It is interesting to note that the t-shirt was popularized by soldiers, as it is a piece of clothing that makes it rather difficult to identify hierarchies visually. The American soldier with a t-shirt on his muscular body turned into an icon, and some commercial photos of the US Army from World War II look like precursors of Rambo. US soldiers still wear t-shirts as an official part of their uniform, and the US Army is apparently working on what they call »smart shirts,« t-shirts that include sensors that send the health data of the soldier wearing them to a computer so that, far away from the battlefield, it can be determined when soldiers are wounded and what can be done for them medically. This might sound like science fiction, but it is already possible to produce t-shirts of this kind.

After World War II, the t-shirt GIs had done so much to popularize became one of the symbols of the victorious United States and, therefore, a symbol of what was seen as »the American way of life«: social equality (or the chance to get it), modernity and mobility. However, the t-shirt was not yet considered socially acceptable clothing in either the US or in Europe. Before the late 1960s university students would not dream of wearing t-shirts to their lectures. It was film, especially Hollywood, that created the image of the t-shirt as the clothing of rebellious youth: Marlon Brando (*A Streetcar Named Desire*, 1951) and James Dean (*Rebel Without a Cause*, 1955) personified this new male role model, presenting the young adult male body as an object of virility and desire. It took much longer for female actors to be shown wearing t-shirts in film. The best-known early example is in *À bout de souffle* (*Breathless*, 1960), where Jean Seberg dons a t-shirt emblazoned with the *New York Herald Tribune* logo, marking her clearly as an American girl in Paris. This was just the beginning of publicly displaying the sparsely covered female form that has since become de rigueur in movies and commercials of all kinds. Not very much the later, the bikini, which was in the beginning highly disputed, began gaining in popularity.[5] It was its appearance in popular movies that made the t-shirt increasingly acceptable, and after a while, American colleges and universities began to produce their own t-shirts, emblazoned with the name of the institution and usually also its crest. These shirts began to replace the more formal blazers, which usually had the logo of the university embroidered on the left side of the breast.

Hollywood was also among the first to use t-shirts as a medium for advertising. In 1939 MGM produced the first t-shirts that were merchandising products for a specific movie: *The Wizard of Oz*. Politics realized rather early on the potential of the t-shirt as a medium. In 1948, t-shirts with election slogans made their first appearance in an American election campaign. Republican Thomas E. Dewey's supporters sported t-shirts with the slogan »Dew it with Dewey.« As it turned out, they weren't able to do enough to stop Harry S. Truman, but the t-shirt as political commercial was established. Among the best-known election t-shirts is that from 1960 with a portrait of John F. Kennedy in front of the stars and stripes and the simple but effective slogan: »Kennedy for President.« T-shirts are an ideal medium for expressing political statements. You do not need to rent billboards or pay for newspaper advertisements; you simply produce some shirts with a catchy slogan and try to get them shown in public or via electronic media. NGOs like Amnesty International regularly produce t-shirts in support of political action. But as they are easy and cheap to produce, also individuals can use slogan-shirts to express their political convictions. Maybe sometimes these convictions can be changed as easily as one changes a shirt, but that is another matter.

Illustration 2: Protester, Vienna, Stephansplatz, 23 July 2004.

How quickly and how globally political expressions can be communicated by wearing t-shirts can be seen with the shirts expressing comments on 9/11.

Different kinds of t-shirts were available for sale three or four days after the terrorist attacks with slogans like »America under attack« or with the well-known New York logo »I love NY« (with the red heart replacing the word »love«) changed into »I love N.Y. more than ever.« Images of Osama Bin Laden, and later Saddam Hussein, in the form of »wanted circulars« and more and more military images also became popular. T-shirts with the portrait of Bin Laden were also worn in some Islamic countries, but with a clearly different message in support of Bin Laden. This can be called a strange case of (controversial but unmistakable) communication between cultures, using the same sign to convey a different meaning and using t-shirts as a global medium of communication. It is said that at some markets in Pakistan, one could buy both versions of the shirt (anti- and pro-Bin Laden), which expresses clearly the in-between situation of that country.

The possibility to communicate political stances is only one part of the t-shirt phenomenon. In the 1960s, the t-shirt began to replace the regular button-shirt and the female blouse in the West, a sign that the clear boundaries between business life and private life were beginning to blur. T-shirts, then, can be and are used to situate oneself in society and express one's identity, or what one wants to be understood as one's identity:

> The self is constituted through recognition [...]. We have here a process of projection and dramatization; the self makes itself what it will then be able to recognize itself as (or in). This occurs through the appropriation of such images or roles as happen to be available; through appropriating, rehearsing and thereby projecting such images, through trying them out (or on), the self comes to know itself as the sort of self it has been privileged to become. Hence theatricality, as this mode of imaginative appropriation and construction, is the process whereby the self can become a fluid, changing, yet continuous creation, carrying the multiple burden of modernity. (Jervis 27)

As the example of the t-shirt shows clearly, these »self-constructions« of personal and social identity are often made in pre-formulated ways. Images or slogans may be drawn on to express personal opinions or maybe individual memories, but they themselves are common and far from unique. Everyone knows the »I love«-hearts (that can become accommodated to every place in the world); the smileys; the tongue of the Rolling Stones; the superman »S«; camouflage shirts; Alberto Korda's portrait of Ché Guevara, which is said to be the most-copied image of the twentieth century; baseball, hockey, football or soccer shirts with the names and/or numbers of sports-heroes; and the many well-known commercial trade-logos. This intermingling of global uniformity and linkage to one specific place is illustrated in exemplary fashion by the much imitated Hard Rock Café t-shirts, which very successfully connect a global im-

Illustration 3: House of Blues, New Orleans

age with both a local and a personal meaning. Of course, such success lends itself to caricature. In the Croatian city of Dubrovnik, I once saw a shirt with a fake »Hard Rock Café« logo and the slogan »There is no Hard Rock Café in Dubrovnik.«

In the 1970s, designers began to print their names and/or design logos on the fronts or backs of t-shirts they produced instead of just on the label inside in the neck. Dolce & Gabbana took on this idea in a playful manner, crossing out their own »D&G« logo and writing »no logo« under it in reference to the critical »no logo« movement. T-shirts with commercial slogans or design logos of companies of any kind are an interesting phenomenon, as these shirts turn the person wearing them (who has paid for the privilege) into a living advertisement for these companies, which in turn, of course, generates reactions in the form of various kinds of faked logos. A typical example is the image shown below of the McDonalds logo and Lenin, which I found in Budapest in the 1990s and which clearly expresses the cultural situation of Hungary in that period of transition between late communism and rising consumerism.

T-shirts are used as merchandising products not only by companies but by

Illustration 4: McLenin's. The Taste of Socialism

nearly every kind of economic institution. Museums, for example, now count on the t-shirts available in their gift shops to supplement what they can earn in admissions. About three quarters of all merchandising products sold worldwide – by museums, rock-bands, concert halls, theatres, restaurants, shops, film companies, TV networks, universities, newspapers, illustrated journals and so on – are t-shirts. This overlaps with another wide field: tourism. T-shirts can be used to demonstrate publicly the places one has visited, as well as representing personal memories or gifts you might bring from a trip. Most of the images are rather standardized and simple, showing typical and popular signifiers of the place in question. They can also deal with single events or cultural practices of tourists, and sometimes one can find more complicated expressions or even ironic comments on public images, as in the illustration below that is rather popular among tourists visiting Austria, playing with the common mistake of confusing »Austria« with »Australia.« As is evident from the two examples shown here, most – but not all – of the t-shirts for tourists do not feature the language of the specific country in which they are being sold. English is clearly

Illustration 5: No Kangaroos in Austria

the leading language of our globalized world: even the word itself – »t-shirt« – is normally used in most languages in its original English form and not translated.

Another large arena of the t-shirt phenomenon is sports. Fans of different teams wear »their« team's shirts and colors more or less like a uniform, clearly marking their belonging to a very specific group of people and differentiating themselves from the opponents. Rather popular are shirts that are direct copies of the shirts the players themselves wear, with the number and the name of a specific player on the backside. Of course, also shirts of national-teams can be bought. They are real cash-cows for the companies that produce sportswear.

When France won the soccer world cup in 1998, within one year Adidas had sold half a million shirts of the French national team in France alone – and of course, many fakes and imitations were sold too.

Another economically important field is that of children's wear.[6] T-shirts are ideal for children, as they are practical, easy to clean and more or less inexpensive. Especially in this area, the tendency to worldwide uniformity is evident. When one of the major film companies, e.g. Disney, launches a new movie for children, merchandising products, especially t-shirts, are sold all over the world. This is not only a children's phenomenon, of course, but as children grow out of their clothes and cannot wear the same shirt for a longer period, they are a huge market, one that longs for actuality without any sentiment of nostalgia and one very much subject to peer pressure. It may be Disney, the Simpsons, Pokemon, Japanese Mangas, Harry Potter or the Lord of the Rings – the motifs are the same all over the world, but signify differently given different local histories and cultures.

In focusing on t-shirts, a strict divide cannot be drawn between art and fashion. One might think of the function of different kinds of t-shirts (e.g. net-shirts) in the gay communities or of street-wear that is worn by urban youth groups and gangs, expressing a feeling and creating an image of a specific, and adventurous, urban way of living – a style that in the meantime has greatly influenced *haute couture*. What is the difference when American artist Jenny Holzer uses t-shirts as surfaces for statements – that she calls »truisms« – and couturier Sonja Rykiel does exactly the same? When painter Navan Ravanchaikul from Thailand produces t-shirts and sends people wearing those shirts in cabs all around and calls that wandering exhibitions, and couturiers produce personal t-shirts as invitations to *prêt-a-porter*? When artists produce t-shirts in limited editions that seem to be unique pieces, and couturiers do exactly the same, calling them customized shirts? One might as well think of the Benetton phenomenon, which seems to indicate that the borders between art and couture have all but vanished.

A rather funny example of the use of t-shirts in an art project is the image shown below. When some years ago the city hall of the Austrian city of Graz, the capital of the province of Styria, was restored, paintings by Manfred Deix, a very popular caricaturist in Austria, were placed on the covered façade, especially there, where the windows were, using a faked logo of »Microsoft Windows« (written »windous,« the way it would be pronounced in strong Styrian dialect). The motifs of the paintings were also available on t-shirts, which were only sold during the period of restoration. It is no longer possible to buy them regularly now, which makes these shirts collectors' items. The most popular of the motifs is that one that was placed in the middle of the façade, showing actor, nowadays politician, Arnold Schwarzenegger, who comes from a small village near Graz, in – of course extremely overdone – old-fashioned Styrian traditional costume.

Illustration 8: Deixcroft Windous: Arnie @ Planet Grazywood

To sum up, t-shirts are an important technology of self-expression in the global landscape: maybe strictly individual, but most times in a pre-formulated way. The images and slogans are countless.[7] They might try to provoke or to attract, to comment on actual political events or social and cultural habits, to express romantic feelings or bad taste, they might be easy to understand or rather enigmatic. Some of them are more, some of them less intelligent, others are sheer nonsense, and some so brutal or disgusting that we might wonder what kind of person would want to wear them. But whatever the message, in the end, as Charlotte Brunel notes, t-shirts communicate.

NOTES

1 Quotations from this book are my translations from the German edition.

2 See Hollander.

3 See also Wittkop-Ménardeu.

4 For the history of male underwear, see Engel.

5 See Berger. For the origins of the name »bikini,« see Kramer. Bikini Island in the Pacific was used by the US Army as a testing area for the first two nuclear bombs tested after the end of World War II in the summer of 1946. That summer the French couturier Louis Réard presented the bikini as a new item of female bathing attire and named it after the nuclear test-area.

6 See Weber-Kellermann.

7 See Walters and Walters and Fletcher.

REFERENCES

Berger, Beate. *Bikini. Eine Enthüllungsgeschichte*. Hamburg: marebuchverlag, 2004.

Brunel, Charlotte. *T-shirt*. Vienna: Christian Brandstaetter Verlag, 2003 [Original: *T-shirt*, Paris: Assouline 2002].

Engel, Birgit. *Die Unterhose*. Berlin: Feierabend Verlag, 2003.

Hollander, Anne. *Sex and Suits*. New York: Alfred A. Knopf, 1995.

Jervis, John. *Exploring the Modern. Patterns of Western Culture and Civilization*, Oxford–Malden/MA.: Blackwell Publishers, 1998.

Kramer, Fritz W. *Bikini. Atomares Testgebiet im Pazifik*. Berlin: Verlag Klaus Wagenbach, 2000.

Walters, Helen and Tim Fletcher. *100% Cotton. T-Shirt Graphics*. New York: Watson & Guptill Publications, 2001.

Walters, Helen. *200% Cotton. New T-Shirt Graphics.* New York: Harper Design, 2004.

Weber-Kellermann, Ingeborg. *Der Kinder neue Kleider. Zweihundert Jahre deutsche Kindermoden.* Frankfurt am Main: Suhrkamp Verlag, 1985.

Wittkop-Ménardeu, Gabrielle. *Unsere Kleidung. Aus der Geschichte der Mode bis zum Jahr 1939*, Frankfurt am Main: Insel Verlag, 1985.

Wyss, Beat. *Die Welt als T-shirt. Zur Ästhetik und Geschichte der Medien*, Köln: Dumont, 1997.

Probe II:

The Becoming of Central Europe

The Pleasures (and Problems) of »Merely Circulating«: The Vienna Philharmonic's *Neujahrskonzerte* and their American Reception

MICHAEL SAFFLE

> *The purpose of anyone who writes the history of any large epoch must necessarily be to impose a pattern on events, or at least to discover a pattern, and for that purpose a sound general theory, or even an instinctive grasp of probability, might be more useful than a mountain of learning. A history constructed imaginatively would never be right about any single event, but it might come nearer to essential truth than a mere compilation of names and dates in which no one statement was demonstrably untrue.*
>
> *George Orwell*[1]

When »floodgates« – the theme of the present volume – was announced in the form of a call for papers in early 2004, I thought about several subjects for a publishable presentation that would address »the instability of cultural flows that are produced by media and how it affects whether and how they are reterritorialized into stable spatialized forms of identification and imaginings of culture.«[2] To put it more directly, I thought about preparing yet another paper devoted to American perceptions of musical Central Europe, especially *Musikstadt Wien*: a cultural construct with a history all its own (Szabó-Knotik). How, though, could I »do« musical Vienna and North America again, this time in terms of »media« issues, especially after having devoted all or parts of three previous articles to movies, the musical-comedy stage, the novel, and the daily press? The answer that came to me – the annual *Neujahrskonzert* (or New Year's Day Concert; hereafter »Concert«), broadcast each year around the world by the Vienna Philharmonic – seemed both plausible in terms of research opportunities and intriguing in terms of potentially emergent patterns of acceptance and »imagining.«

There was a problem, though. No one seemed to have written anything especially interesting about the Concerts, at least not insofar as »interesting« functions as a signifier for a great many postmodern theorists. What already existed in print or online dealt almost exclusively with the history of the Concerts, their programs, and their relationship to political or cultural aspects of Austria's past rather than her present. A search for articles dealing with Vienna, the Philharmonic, and the *Neujahrskonzerte* in such databases as *The Music Index* and *RILM* turned up almost nothing »critical.« Nor did a longer, more probing Google search. I consulted print copies of *The Reader's Guide to Periodical Literature* for 1949-1975 and *The Music Index* for 1949-1978. I searched hard-copy indexes for the *New York Times*, the *Los Angeles Times*, and several other newspapers. Finally, I leafed through more than fifty general and specialized histories of radio and television programming in the United States. I also managed to locate lots of digital ads for CDs as well as a smaller number of advertisements and reviews of Concert videotapes and DVDs. Only the *Reader's Guide* and *New York Times* yielded anything more substantial.

Small wonder, then, that Orwell's suggestions (quoted above) struck me as useful: I would write a history of the Concerts and their American media »presence« and reception out of... well, out of the few facts I could find, as well as what I hoped would be considered »sound general theory.« At worst, I thought, »an instinctive grasp of probability« might come to my rescue. After all, I myself am an American. I already knew something about America's attitudes toward musical Central Europe, especially Vienna. I also understood something of how »imagined communities« have come to be constructed and peopled in various cross-cultural circumstances. And, during the course of my preliminary researches, I did my best to learn what I could about the Concerts themselves: their origins, evolution, personnel, programs, and dissemination – the last, especially, by means of radio, sound recordings, and television. As I read, reflected, and wrote, I chose Wallace Stevens's poem »The Pleasures of Merely

Circulating« (Stevens 54) as a metaphor both for the Concerts themselves and for their American reception of yesteryear and yesterday.

§

Probably no annual musical event is enjoyed by more people around the world than the *Neujahrskonzerte* presented by the Vienna Philharmonic Orchestra. Inspired or anticipated by Austrian musical events of the past 160-odd years, the Concerts actually began as an annual tradition near the end of the 1930s. As early as the late 1950s they were broadcast via radio throughout Western Europe. During the 1960s and 1970s they also began to be televised in Albania, Ecuador, Jordan, Mexico, Singapore, the Soviet Union, and some two dozen other countries. Since 1985 they have become a staple of American holiday television.[3] In addition, individual Concerts have been preserved on long-playing phonorecords, CDs, DVDs, and videotape. Available for over a decade through more conventional retail outlets and mail-order houses, recent Concert recordings are increasingly advertised, reviewed, and sold online.

Historians of the *Neujahrskonzerte* occasionally look as far back as the 1840s, when Johann Strauss, Sr., mounted the podium at »his favourite venue, ›Sperl‹« in Leopoldstadt, to conduct a »combined New Year's Eve – New Year's Day concert«: one that »started late on December 31st and ended in the morning hours of January 1st« (Dieman & Trumler 167). This event and others like it anticipated two important aspects of today's Concerts: the foregrounding of music composed by Strauss family members, and a »well-defined« ritual that, virtually from its inception, incorporated such patriotic pieces as Haydn's imperial anthem. Another date given for the beginning of the Concerts, at least insofar as both the Vienna Philharmonic as an organization and »light classical« programming are concerned, is 6 April 1902, when Joseph Hellmesberger, Jr., led the orchestra in a concert of works by Otto Nicolai. Like a few of the subsequent New Year's Concerts, this event was »extraordinary« (*außerordentlich*) in that it was advertised in addition to, rather than as part of, the Philharmonic's regular season (Hellsberg 1992: 568). Still another date is 1921, when the Philharmonic played *An der schönen blauen Donau* (»On the Beautiful Blue Danube,« better known as the »Blue Danube« waltz) at the dedication of the *Strauss-Denkmal* in Vienna's *Stadtpark* (Weigel 102). Yet another is 1925, when a series of Viennese orchestral concerts celebrated the centenary of Johann Strauss, Jr.'s birth (Blaukopf & Blaukopf 1992: 308). And another: 11 August 1929, when Clemens Krauss, then recently appointed director of the *Wiener Staatsoper* (the Philharmonic's parent organization), conducted as part of that year's Salzburg Festival a program that included the Overtures to *Der Zigeunerbaron* and *Die Fledermaus* as well as the »Blue Danube« waltz (Blaukopf & Blaukopf 1992: 308-9).

Officially, however, »New Year« Philharmonic performances were launched only in 1939 when that orchestra issued a »public invitation to an ›Extraordinary Concert‹« scheduled for 11:30 a.m. on Sunday, 31 December of that year, in the *Musikverein*. (A public dress rehearsal was held at 3 p.m. the previous afternoon; to this day, Concert »rehearsals« and »previews« are heavily attended.) Krauss selected and conducted a program consisting exclusively of Strauss family chestnuts: among them, the *Pizzicato-Polka* composed by brothers Johann, Jr., and Joseph (Dieman & Trumler 168), together with Johann Jr.'s *Fledermaus* overture and *G'schichten aus dem Wienerwald*. World War II soon put a stop to such seemingly frivolous, if heartwarming, events, but New Year's performances began again on 1 January 1946 and have continued, more or less without interruption, to the present. Krauss himself, however, conducted for the last time in 1953.

It was during the 1950s that technology became an overwhelmingly important aspect of the *Neujahrskonzert* tradition. From 1955 to 1979, Willi Boskovsky conducted every Concert, most or all of which were broadcast on Austrian radio (Hellsberg 1991). The 1959 Concert was the first broadcast on television; viewers in Austria, Belgium, Britain, France, West Germany, and other European nations watched it on their black-and-white sets. Ten years later, the 1969 Concert was broadcast in color; according to one source, it was this broadcast that launched color-TV programming in Austria (Dieman & Trumler 166). In 1961, dancers »accompanied« the orchestra for the first time on television, and within a few years dance numbers were being choreographed, rehearsed, and filmed months in advance of New Year's Day, often out-of-doors and in such idyllic summer locations as the Prater and the Vienna Woods. From 1980 to 1986 the Concerts were conducted by Lorin Maazel; since then Claudio Abbado, Herbert von Karajan, Zubin Mehta, Riccardo Muti, and other celebrities have appeared as Concert conductors. Radio and television audiences for the Concerts grew for several decades at an almost exponential rate. The 1991 *Neujahrskonzert*, for example – the first Concert broadcast in the People's Republic of China (Blaukopf & Blaukopf 1992: 310) – added some 200 million viewer-listeners to the 400 million Europeans, Australians, Japanese, and North and South Americans that had watched many or all of the 1980s broadcasts (Blaukopf & Blaukopf 1986: 275).

The *Neujahrskonzerte* remain »extraordinary« in several senses. Although a »normal« feature of Viennese musical life, each Concert requires what some might consider »abnormal« energy and enthusiasm on the part of would-be attendees. Until very recently, requests for tickets to both the following year's Concert and one or more »preview performances« were accepted by mail or delivery in person on *just one day each year*: generally, the first regular workday in January. For the 2003 Concert, that day was 2 January 2002. For the 2006 Concert, however, the rules were changed and requests were accepted between 2-27 January 2005. Traditionally, requests have been limited to four tickets for each »foreign« order and two tickets for each order by Austrian citizens or residents. After requests are received, ticket recipients are chosen by means of a lottery. Prices, always steep, continue to climb. The cost of admission to the 2006 Concert, and to a second New Year's Eve performance, ranges from 170 to 570 Euros; admission to the 30 December 2005 »preview performance« costs between 130 and 380 Euros.[4] By contrast, 2002 ticket prices ranged from 109 to 436 Euros for the 2003 Concert, and from 87 to 291 Euros for the »preliminary« 30 and 31 December 2002 performances.

More than a few Viennese music historians, including specialists in the Philharmonic's activities and reception, have foregrounded the musical character of the various Concerts – in part, because the Philharmonic »traditionally« plays only more serious or »classical« fare; in part, too, because the music of Lanner, Nicolai, Offenbach, and the Strauss family remains especially popular with audiences in Vienna and around the world.[5] For most of these historians, as well as for a great many concert-goers, the works of Johann Strauss, Jr., together with such patriotic numbers as Haydn's anthem[6] and the Radetzky-Marsch, »are« Vienna.[7] During the 1940s and early 1950s, in fact, the post-war Concerts occasionally functioned as unofficial political memorials (*Denkmäler*) or celebrations. Johann Strauss, Jr.'s Russischer Marsch was performed at the 1941 Concert. On New Year's Day 1941, the Non-Aggression Pact signed by Hitler (on behalf of »greater Germany,« including Austria) and Stalin was still »in force.« On 22 June 1941, Hitler's army invaded Stalin's Russia (Dieman & Trumler 168).[8] The 1955 Concert opened with Josef Strauss's *Mein Lebenslauf ist Lieb' und Lust*, in honor of Austria's then-recent declaration of sovereign neutrality (Dieman & Trumler 168, 201).

Small wonder, therefore, that dozens of Germanophone musicologists have referred in their monographs, encyclopedia entries, and professional periodical articles to the Concerts (e.g., Biba; Boese and Rottensteiner; Grasberger 1963; Jerger; Kralik; and Strasser). More than a few experts have devoted whole chapters to Concert history and programming (e.g., Blaukopf & Blaukopf 1986: 275-80; Blaukopf & Blaukopf 1992: 308-11; Hellsberg 1992: 568-73; and Weigel 98-106). And several authors have devoted entire volumes to the Concerts in terms of their antecedents, character, evolution, and influence (Dieman & Trauber; and Dieman-Dichtl). On one occasion, the confluence of »Philharmonic,« »Concert,« and »Strauss« resulted in an official exhibition catalogue (e.g., Grasberger 1965). For the most part, however, the Concerts have in some sense been trivialized or taken for granted, especially as »merely« delightful manifestations of Vienna's artistic, economic, and political past.

Remarkably, the Concerts themselves – which is to say, their music, production, presentation, and reception – have been almost entirely overlooked or ignored by *America*'s musical-critical community. Like cattle skulls in the woods, they have so far kept their secrets. I do not mean to imply that Americans are uninterested in Vienna and its cultural heritage. In 1986, to cite but one example, more than a dozen popular American magazines – *Harper's Bazaar*, *Horizon*, *Time*, and even *Better Homes and Gardens* – ran reviews of an exhibition of Viennese *Jugendstil* and Expressionist works mounted at Manhattan's Museum of Modern Art. The *Neujahrskonzerte*, however, were scarcely mentioned in 1986 (but see Fantel) – and this, even though that year witnessed only the second televised Concert broadcast in the United States.

Nor do I mean to imply that my nation's critical silence necessarily contradicts or calls into question the overwhelming popularity of the Concerts as holiday TV fare. For literally millions of Americans, the Concerts have proven themselves wholesome and diverting. Perhaps, in fact, they have attracted so little critical attention because they have become *too* familiar. Like the Rose Bowl Parades, the Concerts simply show up annually on our New Year's Day TV screens. Each year they charm us for a moment or two. Then – for everyone, save those enthusiasts who collect them in recorded form – they pass into electronic oblivion. Unlike the football games that mostly precede them,[9] no one »wins« or »loses« listening to the Concerts. They are, simply, *there*. And (thank goodness!) they're painless to look at and listen to. After all, for a great many revelers New Year's Day is the quintessential »morning after the night before.« This fact, too – that hangovers and heavy meals dull the mind as well as the body – has unquestionably contributed its share to an absence of thoughtful *Neujahrskonzert* critiques.

The stability of the Philharmonic's programming also has something to do with the Concerts' critical invisibility; so does the light-hearted character of the music itself. Although works by Mozart, Schubert, and other »heavies« are played from time to time, and although »obscure« composers such as Joseph Lanner are occasionally highlighted – 2001 was Lanner's »year« – musical programming varies little from Concert broadcast to Concert broadcast. As do the visuals. Shots of the gilded caryatids that support the balconies of the *Gesellschaft der Musikfreunde*'s Great Hall, shots of ladies in elegant gowns and gentlemen in suits and ties, shots of the orchestra, and lots and lots and lots of Strauss: this is the standard sequence of events, interwoven with spoken commentary and a little dancing. Always the orchestra plays a couple of novelty numbers, the *Pizzicato-Polka* and *Tritsch-Tratsch »Polka schnell«* among them. And, always, they end with the *Radetzky-Marsch*.

The phenomenon of the televised *Neujahrskonzerte* and their simultaneous critical invisibility – which is to say, the phenomenon of an immensely popular »foreign« musical ritual at one and the same time embraced and ignored by

»natives,« even by trained observers eager to comment on virtually anything »entertaining« or »cultural« – seems to me worthy of note. If silence implies consent, as it does in certain legal situations, then the silent acceptance of the Concerts by American commentators and audiences would seem to imply consent insofar as certain stereotypical North American responses to Vienna, her music, and her citizens are concerned. How is it (I ask rhetorically) that »we« agree to enjoy the Concerts without in any significant way (re)evaluating them? Because (I answer rhetorically), »we« already understand what »they« are, who listens to »them« and why, and what »their« overall image signifies for »us,« as Americans.

Consider the following hypothetical assertions, which I am forced to construct imaginatively (as Orwell suggested) in the absence of that »compilation of names and dates« I would prefer to invoke. A great many of the men and women who watch, or at least glance at, each year's Concert broadcast may not know – most of them, I'm pretty sure, neither know nor care – what Haydn's hymn meant to Austrians. These same men and women have probably never heard of *The Merry Widow* or read Marcia Davenport's *Of Lena Geyer*. Many of them, however, can identify the »Blue Danube« waltz, just as they can recognize several of Mozart's most familiar works, because, like millions of other Americans, they – *we* – have seen *2001: A Space Odyssey* and *Amadeus*. Our perceptions of Central Europe have been shaped more by the constructed images of American popular culture, by movies like *The Sound of Music*, than by actual visits to Salzburg or Graz. Just as our grandparents and great-grandparents formed their own impressions of Vienna from Lehár's operettas or Davenport's fiction. Having already understood (which is to say, imagined) what Vienna looks like and sounds like, we third-millennium Americans – like our ancestors – evidently believe we need look and listen no further.

In several previous publications, I have written about the pervasive images that we Americans have constructed for ourselves of *Musikstadt Wien*. These images have evolved over time; to some extent they have occasionally been corrected and even contradicted, but they have never altogether been abandoned. Deeply embedded in our collective cultural psyche are notions of Vienna – and, by extension, of Austria and Central Europe – as a Land That Time Forgot, an imagined community of simultaneously patrician and plebeian *fin-de-siècle* delights. These notions and imaginings are inescapably bound up, as I have already explained, with *The Merry Widow*, *Of Lena Geyer*, and *The Sound of Music* (Saffle 2001). Even *New York Times* reporters merely dented those images between the World Wars (Saffle 2002). After World War II, the tours and recordings of the Trapp Family Singers reinforced a great many preconceptions throughout Canada and the United States. Today, the Trapp Family Lodge in Stowe, Vermont, memorializes a largely imagined Austria in concrete, stone, and wood (Saffle 2003).

The *Neujahrskonzerte* have done little to disturb our collective convictions. Nor has the Vienna Philharmonic, at least insofar as its televised persona is concerned. Of everything retro about Austria's capital, nothing, perhaps, is more retro than the New Year's Day Concerts. Socially elitist they have always been, and elitist they remain. At the same time, the audience that attends the Concerts seems – at least momentarily, at least on TV – to constitute itself as a temporary democracy, one devoted to celebration. Somehow, too, in spite of restrictions governing ticket purchases, the same few wealthy big shots (or their clones) – all of them, for the time being, delightful (and delighted) members of that temporary democracy – are on hand, each year, snug in seats close to the conductor's podium, as the electronic curtain once again goes up around the world, and something like a billion viewers[10] gaze upon comfortable, mostly Austrian faces while sonic reminiscences of a vanished Operetta Empire swirl and froth about those viewers' eyes and ears.

Either it is true that, each year, we Americans swallow the Concerts as one might swallow a bromide, largely because each Concert merely reinforces our national predisposition in favor of fun-loving, music-making Central Europeans. Either that, or an awful lot of critics have missed out on a golden opportunity to Go Public with novel explanations as to why so many of our fellow citizens listen every New Year's Day to music they otherwise mostly or entirely eschew. Hypotheses aside, what has happened – and what seems mostly still to be happening in the United States – is that the *Neujahrskonzerte* »merely circulate.« Vienna's musical clouds not only go round and round our heads; they go round and round themselves. They are, in some fundamental sense, ignored. *Invisible*. Even as, or perhaps *because*, they are enjoyed.

§

Only during the last few years has this situation slowly begun to change. Thoughtful articles have recently appeared in print, their contents insightful and even confrontational rather than merely complimentary. Instead of »classical sounds« (or only such sounds), a grappling with issues of ethnicity, politics, and gender. An ever more sophisticated interest in, and concern with, the changing roles played by radio, sound recording, TV, and digital forms of communication in Western culture. As more of us question our assumptions about musical Vienna, we grow more aware of the *Neujahrskonzerte* as a remarkable cultural phenomenon. The drummers, it would seem, are at long last starting to rumble something out of their drums.

Concert radio broadcasts were all but ignored by the »serious« American press throughout the 1960s and 1970s, although a great many ephemeral publications provided their readers with programming schedules and holiday entertainment announcements.[11] When the Concerts began to be televised in the United States, however, the situation began to change and audience response to diversify. This did not happen overnight, however. Nor has it happened often.

True, the *New York Times* and *Village Voice* acknowledged the first few years' broadcasts in terms of disruption as well as stability. In 1985, for example, the *Times* commented on »the first United States telecast of the New Year's Day concert of the Vienna Philharmonic« in part by observing that two years earlier, in 1983, Viennese demonstrators had »demanded to know why Lorin Maazel, an American, was conducting their revered orchestra.«[12] The 1984 Concert news from Austria's capital was even juicier: two men, the *Times* reported, had »leaped« that year in front of »the Viennese audience in the historic *Musikverein*, stripped off their clothing and unfurled banners for the homosexual-rights movement.«[13] As certain Concerts were transformed into sites of public contention, their newsworthiness increased. On the other hand, the *Times* concluded its 1985 report by observing that »yestterday [*sic*], as the reassuring presence of Walter Cronkite, the host, appeared« in Vienna's television studios to introduce that year's American Concert broadcast, »station executives drew a small sigh of relief. And, for the next 90 minutes, the *main Austrian export for the last two centuries* prevailed.«[14]

Some three weeks later, Leighton Kerner published in the *Village Voice* a leisurely but largely uncritical description of the same *Neujahrskonzert* as a Central European musical confection. Confessing that he had »no chance« of attending the Concert proper, Kerner acknowledged himself lucky when he received a »consolation prize [...] in the form of tickets [for his wife and himself] to a Saturday morning preview, with the entire program, indispensable encores and all, played for an audience mainly of Austrian army personnel and a small contingent of civilians.« Kerner then turned, somewhat fulsomely, to the music and its stereotypical cultural associations:

> To hear the Vienna Philharmonic in its home auditorium, the elaborately chandeliered and gold-decorated Great Hall of the *Musikverein* (a short name for the German equivalent of Society of Friends of Music) is to hear orchestral sound at its warmest and plummiest, and when the subject is the waltzes, polkas, and marches of the Johann Strauss family – at this concert Johann father and son, plus sons Eduard and Joseph – warmth and plumminess are enhanced with a buoyancy and relaxed lilt that makes it very hard not to kick your chair aside and dance – even an elephant like me.

After dealing with conductor Maazel and summarizing the Philharmonic's relationship to the *Staatsoper*, Kerner finally got around to comparing real life with virtual televised reality, adding that »air-time viewers« may prefer to »switch off the TV sound, turn on their stereo FM, and get a digital transmission«:

> What American viewers saw [on 1 January 1985], by the way, was more than Europeans got on their TV sets, but not quite what they had been led to expect. First of all, only the second half of the concert was televised – anywhere. Radio carried the whole program live, at least in Austria, but the TV and radio transmission, hosted by Walter Cronkite, began after the *Musikverein* intermission, with the one non-Strauss piece on

> the program, a zippy performance of Berlioz's *Roman Carnival Overture.* […] Vienna did *not* get Cronkite's brief, history-catching reminiscences-cum-film-clips as a working reporter in that city, or his superbly relevant comments on earlier history, political as well as musical. And Europe didn't get a handsome, evocative epilogue evidently concocted by Channel 13's executive producer, Peter Weinberg [… that] reviewed the whole quarter-century of New Year's Day telecasts. (Kerner)[15]

In other words, the TV medium itself constitutes much of Kerner's message. Certainly more of it than the significance of the Strauss family's legacy, or the politics of 1980s Austria, or even Maazel's musicianship. Instead, an account of what a great many music-lovers have come to consider one of television's most newsworthy attributes: the transformation of »hot« living reality into a »cool,« electronically mediated experience. And a little history, courtesy of Cronkite.

In 1986 the *Times* ran a longer article about that year's Concert broadcast, the second televised in the United States. This time the emphasis was almost entirely electronic. Entitled »Technical Innovations in Vienna TV Concert,« author Hans Fantel's observations appeared in the prestigious first section of the paper and opened with the following paragraphs:

> Because of the scope of its distribution, the annual New Year's celebration of the Vienna Philharmonic has traditionally served as a proving ground for technical innovations in long-distance transmission. This year, American viewers will enjoy a new advantage. At least 26 PBS stations in major cities are now equipped to transmit stereo sound and can take advantage of the stereo signal being supplied from Europe.
>
> WNET in New York will be broadcasting in stereo, and viewers with suitably equipped television sets will be able to receive the broadcast in this superior mode. Others can tune their stereo systems to FM station WNCN (104.3 Mhz), which will simultaneously carry the audio portion of the broadcast. (Fantel)

Although Fantel mentions »the waltzes of Johann Strauss« and acknowledges that the program as a whole »easily straddles the divisions of Europe and appeals equally to America,« he concentrates on technological issues. Chief among these is the fact that electronics plays a most important role in fostering international communications, as well as localizing a kind of entertainment in terms of Central European traditions and stereotypes (the last, subjects Fantel overlooks or ignores). Everything is perceived through American goggles. Even the »golden hall,« where »the Philharmonic has performed under Brahms, Bruckner, Mahler, and Johann Strauss himself,« is presented by Kerner as a model for America's own engineering success; Boston's »famous Symphony Hall,« it seems, was »patterned« after the *Grosse Saal*'s mathematical proportions.

Throughout the 1990s most American Concert commentators emphasized technological issues over musical or cultural criticism. Old-fashioned aesthetic arguments, often either soft-pedaled or stridently emphasized where *Unterhaltungsmusik* is concerned, all but vanished from sight. (In the case of the Concerts themselves, aesthetics have rarely been foregrounded. Not musicality, but music's role in representing Vienna's *past* traditions, pleasures, and even politics: these things have mostly been given more attention.) Online, however, the story is somewhat different. Pop music and prettiness sell, and web-page ads for »product« often uncritically celebrate the electronically processed sounds of Concert recordings and videos, commodifying them as »listenable« and »enjoyable.« Musicianship and musical knowledge are mostly taken for granted. Thus one CD ad praises Nikolaus Harnoncourt for his leadership of the 2003 Concert, observing that, »for Harnoncourt, European dance forms have always been a family concern.« (Remember: these words were endorsed by iclassics.com; the firm's motto is »raise your standard of listening.«) But which dance forms? Evidently, everybody is supposed to have heard of polkas and waltzes. As for interpretations of particular pieces or passages: forget it! The folks at iclassics.com are in business to make money, not to chat about art.

Even more explicitly commercialized internet pages, among them a sendit.com ad for DVDs of Seiji Ozawa's 2002 broadcast, do little more than mention the Strauss dynasty before honing in on such technological selling points as »Duration: 100 mins,« »Region: 2,« »Presentation: Widescreen,« »Aspect Ratio: 16:9 Wide Screen,« and of course »Cat. No.: DVWPNK02(TDK UK LTD).«[16] Political, cultural, and even musical observations are nowhere to be found. More common are pages that merely announce sales items – in this case, DVDs of the 2001 Concert – in terms of electronic details. »Audio (Dolby Digital 5.1, 448 Kb/s)« is a typical entry. One page provides a complete program for the 2001 Concert broadcast, but restricts »criticism« to a few intriguing observations itemized under the heading »Plot Synopsis.« Among these are: »[P]eople are still wandering into magnificent buildings from a much earlier age to listen to superb music created by and for a much different age« [*sic*]; »These concerts [...] have often produced some wonders, perhaps none more so than the magnificent concert of 1987 under the baton of Herbert von Karajan – generally regarded to be the best«; and »[T]here would probably be a riot in Vienna if the Radetzky March was not played« each year.[17]

On a PBS Great Performances web page, by way of contrast, browsers are invited to »read what our viewers have written« about individual arts events. Here, highly personal responses rather than technological jargon are held up for praise. Unfortunately, however, comments about the *Neujahrskonzerte* are restricted to such platitudes as: »The New Year's celebration concert of the Vienna Philharmonic has been a favorite and key event in our celebration of the New Year for several years«; and »For lack of a better word, I'm hooked on the ›Radetzky March.‹«[18]

Several of these attitudes are also reflected in the only two recent, hard-copy, *critical* discussions of the Concerts and contemporary cultural issues that I have been able to locate: both of them RILM-indexed, both published in specialized musical magazines during the past five years. The first is Tully Potter's review of »Historic New Year's Day Concerts« recordings, which appeared in *The International Classical Record Collector* during the summer of 1999. The second is William Osborne's »Ozawa Conducts the Vienna Philharmonic's New Year's Concert: Tokenism and Public Relations,« which appeared in the *IAWM* (or »International Alliance for Women in Music«) *Journal* in 2002. The former essay is primarily acoustical in its subject-matter, although other concerns receive at least a little attention. The latter is primarily cultural-critical; as peer-reviewed professional work, it occupies an all but unique status among English-language publications dealing with the Concerts.[19]

Potter's »Historical Concerts« piece identifies and describes a number of long-forgotten performances according to their musical and acoustical merits. Praising the »lilt« that Clemens Krauss brought to »this repertoire« (i.e., the Strauss waltzes), for example, Potter goes on to mention that conductor's recordings for HMV in 1929-1931 and for Telefunken during World War II. In Potter's opinion, later Decca releases of recordings by Krauss, themselves originally »made with the benefit of tape and ›ffrr‹ techniques[, ...] stand up well; there is some ›fizz‹ on the strings but the mono image is that of a believable orchestra in a setting of great [sonic? spiritual?] depth« (Potter 82). Potter writes about musical beauty, but primarily – or, at least, more often – in terms of acoustical purity and fidelity, as preserved or damaged by recording processes of various kinds. Having surfed the web for CD and DVD Concert reviews, I found myself conjuring images of digitalized, wide-screen drummers, their hooded rumblings empty of meaning but vibrantly »present« in bursts of high-fidelity sound.

Osborne's take is quite different. Referring to several of his own publications in passing, he criticizes the Philharmonic for its »gender and ethnic uniformity.« The orchestra as an ensemble, he maintains, »has traditionally excluded visible members of ›racial minorities‹ from rank and file membership [...] in the belief that they would destroy the ensemble's image of Austrian authenticity.« Ozawa, Osborne notes, »is keenly aware of his unique situation« as both racial and musical outsider among the Viennese:

> In a news conference before his visit [to Vienna to conduct the 2002 Concert], Ozawa told reporters that he could no longer ignore [political] developments in Austria. He stated: »To me, music must be away from the political movement. I am not thinking of changing my mind about my musical life because of a politician but, again, I must go and see.« (Osborne 21)

Osborne refers to Jörg Haider, the politician in question, as a man who »praised both Hitler and the SS, and has referred to the concentration camps of the Holocaust as ›punishment camps,‹« a man who »campaigns on a politic of anti-foreigner xenophobia« (Osborne 21). Osborne also draws attention to the Philharmonic's unwillingness to hire women musicians, observing that during the five years since the orchestra »voted to end its discrimination, it has hired only one woman who does not play harp – violist Ursula Plaichinger«:

> If the Philharmonic continues to hire [only one] woman [...] every five years it will take 75 years to hire the 15 women necessary to represent even ten percent of the orchestra['s total membership]. Since at least half of those [hypothetical] women would retire within that time, the current rate of employment will never allow women to exceed token levels of about five percent. (Osborne 22)

For two-thirds of the article in question, Osborne devotes himself almost exclusively to political and gender issues. The last third of his observations, however, take up the televised portion of the 2002 Concert *especially* in terms of performance and reception. Change, Osborne observes,

> is slowly becoming apparent. The Philharmonic traditionally performs its waltzes in unique and almost inimitable ways. The players often largely ignore the conductors who lead the New Year's Concerts (such as Muti and Maazel), since most are familiar with these very specific styles. But this year the orchestra followed Ozawa rather closely. [...] Of course, there were some conflicts in works such as the *Wiener Blut* waltz. [...] Perhaps for that reason the television cameras spent most of the time during that work focused on the chandeliers and ceiling murals. (Osborne 22-3)

Those clouds again, going round and round – and located, in this instance, above the heads of audience members, momentarily invisible to TV's all-seeing eye. »Afterwards,« insofar as Osborne is concerned, »the television commentator's remarks about ›free, fresh, authentic and red Viennese blood‹ brought back memories of the orchestra's bizarre ethnic and racial ideologies« (Osborne 23). On the other hand, Osborne acknowledges that Austrian State Television (ORF) briefly foregrounded »an Asian child« during one of the Concert's televised ballet sequences and »also showed some Japanese members in the audience« (Osborne 23).

Osborne seems to suggest that the Philharmonic's members followed Ozawa during the 2002 *Neujahrskonzert* because, in being televised following him, they – as »*the* Philharmonic,« as representatives of *Musikstadt Wien* – would be perceived by audiences around the world as submitting to the artistic leadership of a non-European. Was the broadcast rigged as a demonstration – incomplete, imperfect, but nevertheless deliberate – of a multicultural Vienna? As an international Music City? Intriguing questions!

What I find especially interesting about Osborne's comments concerning gender, politics, and race is that he experiences and reports on the Concert he witnessed largely *in spite of* its history and traditions. He evaluates that »musical« event, together with racial, political, and gender issues, largely in terms of televisual content and reception. Again: the medium is at least part of the message. Perhaps most of it. And again: without radio, various sound recording techniques, and television, there would be no New Year's Concert broadcasts – none, at least, insofar as most of the international musical community is concerned. No possible way of »positioning« them critically, either – again, at least, in terms of their easy-listening contents, conductors' discographies, or multicultural contexts. Which, of course, would mean: even less for musicologists like me to do.

§

If the *Neujahrskonzerte* are beautifully programmed and presented as past music, if they are important in terms of past and present technological and communications developments, and if they reflect certain present social and racial stereotypes that assume variegated forms in various parts of the world, they also raise questions about the evolution of Austrian national identity, just as they raise questions about American perceptions of that identity.

In a critical assessment of »the clichés of *Heimat* Austria,« especially those devolving upon the years between 1955 (when the Second Republic was founded) and 1995 (when Austria entered the European Union), Margarete Lamb-Faffelberger proclaims *The Sound of Music* »the ultimate gift bestowed upon Austria by the United States.« In her opinion, that musical-comedy-turned-movie embodies the spirit of »*Kakania*, an era lost forever« as much as, and possibly more than, many »authentic« Austrian culture products. An imagined world that Lamb-Faffelberger epitomizes primarily in terms of its (re)presentation through music and the other performing arts:

> To play an instrument and sing in the school choir, to listen to Mozart and Beethoven [...] to see the great actors Paula Wessely and Attila Hörbiger at the *Burgtheater*, to attend elegant balls during the Carneval season in Vienna, to indulge in the romantically sweet Sissy-films, and to go hiking and skiing in the Alps like Luis Trenker and Toni Seiler in the enticing *Heimat* films: these are clichés and stereotypes that define modern Austrian identity. (Lamb-Faffelberger 294; italics in the original)

For Lamb-Faffelberger, as for Osborne, »the political rise of Jörg Haider symbolizes the failure in establishing a sound cultural identity in Modern Austria« (Lamb-Faffelberger 297). I agree with her.

I do not, however, altogether accept the notion that the *Neujahrskonzerte* have long been a kind of confidence trick: one way (among many) of hiding Austria's troubled twentieth- and twenty-first-century political and social struggles under a blanket of televisual entertainment. Nevertheless, it seems obvious to me that – like Viennese operettas, like *Of Lena Geyer* and *The Sound of Music* – the Concerts have become, for Americans, more »authentic« than much of recent Austrian history. More authentic too, it seems, for many Austrians. Alternative events may provide amusement for the few; jazzed-up »takes« on Strauss favorites [*verjazzte Strauß-Stücke*], for example, were performed on New Year's Day 2000 by the Vienna Arts Orchestra (Tosic). Such events, however, neither reach the many nor set the record straight. It seems that whether Mrs. Anderson's Swedish baby is, in fact, German or Spanish or whatever (to allude to Stevens's poem) remains – in fact, as well as metaphorically – very much the issue. What *is* »Austria,« anyway, at least in some overarching cultural sense? What have the Concerts contributed to our own, media(ted) definitions, at home and abroad?

Historians of Austrian music, and especially of the Vienna Philharmonic, have implied over and over again that their nation is »brotherly,« that it »reaches out« to the rest of the world. Consider the titles of half a dozen post-World War

II volumes: the Philharmonic is an ambassador [*Botschafter*] to the world, bringing »culture« to other, possibly less enlightened nations (Biba; Boese & Rottensteiner). An orchestra with a »mission« [*Sendung*] (Jerger). A »world orchestra« as well as »the world's orchestra« (Blaukopf & Blaukopf 1992). An ensemble that embraces millions – from Schiller's phrase »O ye millions, I embrace ye« [*Seid umschlungen, Millionen*], as set to music by Beethoven in his Ninth Symphony (Dieman & Trumler). A veritable »democracy of royalty« [*Demokratie der Könige*] (Hellsberg 1992). Certainly the Concerts do nothing to contradict these implications, although they are televised in certain ways so as to emphasize the democracy rather than the royalty (and vice versa). They also exemplify certain Austrian manifestations of sexism and racism. And, if Lamb-Faffelberger is correct, it is the Concerts and similar Austrian culture-products, as well as that quintessentially American cultural product *The Sound of Music*, that define »Austrian identity« today.

As cover-up? That is the question. If we, as Americans, want to know more about the »real« Austria, or want to understand why we would want or need to imagine such an entity as being more than »merely« virtual, we may wish to pay a little more attention to such cultural clichés as the New Year Concerts.[20]

NOTES

1 Quoted from Orwell's review of *The Martyrdom of Man* by Winwood Reade (*The Collected Essays, Journalism & Letters of George Orwell*, ed. Sonia Orwell and Ian Angus; Vol. 4, »In Front of Your Nose, 1945-1950« (Boston: Nonpareil, 1968), 116).

2 Cf. the introduction to this volume.

3 At least one historian, however, claims the Concerts were televised in America as early as 1973-1978 and again in 1981 (Dieman & Trumler 203). Similar disagreements can be found in accounts of Concert radio broadcasts of the 1950s and 1960s.

4 <http://www.wienerphilharmoniker.at/index.php?cccpage=faqs_answers&set_language=en>. Accessed 1 March 2005.

5 Given the prominence of the Concerts in terms of the Philharmonic's world-wide reputation and income, surprisingly little about them appears even in many histories of that

ensemble. Consider Clemens Hellsberg's *Demokratie der Könige: Die Geschichte der Wiener Philharmoniker,* which devotes only six pages to the Concerts (568-73) and ignores in its list of sources (677-8) Dieman & Trumler's book-length account of them – in 1992, the only such account available in English or German.

6 Until 1918 Haydn's melody was used as *Volkshymne* (»people's anthem«) in the monarchy; in 1922 it became the anthem of the Weimar Republic, and in 1929 it became the anthem of the Austrian republic. After the *Anschluss*, new lyrics were adopted, but the melody remained unchanged.

7 A great many web-based references to the Philharmonic, its performances of works by Strauss and his kin, and related issues identify as antecedents of the Concerts events that have nothing directly to do with New Year celebrations. In addition to several of the historic performances mentioned above, for example, these events include the 22 April 1873 premiere by the Philharmonic of *Wiener Blut*, conducted by Johann Strauss, Jr.; a program presented by Strauss and the Philharmonic on 4 November 1873 at the Vienna World Exposition that included the »Blue Danube« waltz; a repeat performance of this last concert on 23 February 1878; and a commemorative concert honoring Strauss on 14 October 1894. See <http://www.viennaonline-ezine.com/features/0202newyear_concert.html>. Accessed 1 March 2005.

8 The *Russischer Marsch* was programmed again »as music« rather than as political commentary only in 1972.

9 Radio and television Concert broadcasts in the United States mostly take place during the evenings of New Year's Days. Some networks, however, rebroadcast previous years' Concerts on New Year's Day mornings or early afternoons.

10 According to one source, »over one *billion* people worldwide« listened to the 2002 Concert as disseminated via radio and television »through about 50 broadcasting corporations« (Osborne 21; italics in the original). Recently, however, Concert broadcasts have been cancelled in China, very much reducing the size of the Philharmonic's potential international audience.

11 Print sources of news are rapidly being supplanted by virtual sources. An example: central California broadcasts of the 2002 Concert, conducted by Seiji Ozawa, were announced during December 2001 not only in local newspapers but online by National Public Radio. In addition to a photograph of Ozawa and a few words praising »the Golden Hall of the *Musikverein* in Vienna, Austria,« one NPR web page identified five

»listener-supported, non-commercial« stations that broadcast the Concert to listeners in Groveland, Sacramento, Stockton, Tahoe City, and surrounding California communities. A second example: Terry Morrow of the Scripps Howard News Service distributed via both print and virtual venues a synopsis of 2004 New Year's entertainment events. Mentioned in this synopsis were MTV's Times Square New Year's Eve TV broadcast as well as a similar celebration hosted by Dick Clark on ABC television, and – erroneously (see note 15 below) – the »fact« that Walter Cronkite would once again host »the annual Vienna Philharmonic New Year's Day concert« (i.e., broadcast): in this instance, at 8 p.m. EST on PBS. See <http://www.csus.edu/npr/images/PressRel/fromvienna.html> as well as <http://www.post-gazette/com/tv/20031231nyeve1231p2.asp>. Both accessed 1 March 2005.

12 »Trans-Atlantic Baton,« *New York Times* (2 January 1985), II, 3:2.

13 Loc. cit.; italics added.

14 Loc. cit.; italics added.

15 In 2003 Cronkite was forced to undergo surgery that made it impossible for him to attend and comment on the 2004 Concert. Instead, he was interviewed by an Associated Press reporter about his enthusiasm for classical music. The interview ends with observations about Cronkite's »unabashedly liberal« newspaper column, in which he »criticizes the Bush administration.« Certainly without meaning to suggest any such thing, the interview as a whole (available online <http://www.sptimes.com/2004/01/01Artsandentertainment/Cronkite_ready_for_an.shtml>) might be construed as mapping a certain brand of musical enthusiasm onto liberal politics. The web page in question was accessed 1 March 2005.

16 <http://www.sendit.com/video/item/7000000068388>. Accessed 1 August 2004.

17 <http://www.michaeldvd.com.au/Reviews/NewYearsConcert2001.html>. Accessed 1 August 2004 and 1 March 2005.

18 <http://www.pbs.org/wnet/gperf/anniversary/comments20.html>. Accessed 1 August 2004.

19 Osborne cites other articles by himself dealing with the Concerts, but they appear only to be accessible online. The articles he wrote for such American newspapers as the *Boston Globe* and *Los Angeles Times*, as well as from the German news magazine *Focus*, are not »Concerts-specific« (Osborne 23).

20 I would like to correct a typographical error in Saffle 2003: the word »Republication« should be »Republican« (p. 87). I would also like to add a very brief amplification to the passage in question: although words such as »mainstream,« »conventional,« and »moral« have, indeed, become opprobrious to many Americans, they may never become obsolete – a point raised by a host of commentators (e.g., Himmelfarb). Finally, I would like to thank Virginia Polytechnic Institute and State University for financial support that enabled me both to attend the »Floodgates« workshop in Toronto between 5-8 August 2004, and to complete the present article.

REFERENCES

Biba, Otto. *Die Wiener Philharmoniker, Botschafter der Musik: Ausstellung im Rahmen des Internationalen Brucknerfestes, Linz 1976.* Linz: Linzer Veranstaltungsgesellschaft, 1976.

Blaukopf, Herta, and Kurt Blaukopf. *Die Wiener Philharmoniker: Wesen, Werden, Wirken eines grossen Orchesters.* Vienna: P. Zsolnay, 1986.

_____. *Die Wiener Philharmoniker: Welt des Orchesters, Orchester der Welt.* Vienna: Löcker, 1992.

Boese, Helmut, and Franz Rottensteiner. *Botschafter der Musik: Die Wiener Philharmoniker.* Vienna and Munich: Österreichischer Bundesverlag, 1967.

Dieman, Kurt, and Gerhard Trumler. *O Ye Millions, I Embrace Ye (Seid umschlungen, Millionen): The New Year's Concerts of the Vienna Philharmonic.* Trans. Edwin Kowanda. Vienna: Österreichischer Bundesverlag, 1983.

Dieman-Dichtl, Kurt. *Wiens goldener Klang: Geschichten um die Wiener Philharmoniker und ihr Neujahrskonzert.* Vienna: Amalthea, 1996.

Fantel, Hans. »Technical Innovations in Vienna TV Concerts,« *The New York Times* 1986.

Grasberger, Franz. *Die Wiener Philharmoniker bei Johann Strauß.* Vienna: Verlag Brüder Rosenbaum, 1963.

_____. *Die Walzer-Dynastie Strauss. Eine Ausstellung zum Neujahrskonzert der Wiener Philharmoniker: Dezember-Jänner 1965-1966.* Vienna: Archiv-Museum, Gesellschaft der Musikfreunde, 1965.

Hellsberg, Clemens. »Der Sonderbotschafter der Wiener Philharmoniker: zum Tod unseres Ehrenmitglieds Willi Boskovsky.« *Musikblätter der Wiener Philharmoniker* 45/8 (April 1991): 270-1.

_____. *Demokratie der Könige: die Geschichte der Wiener Philharmoniker.* Mainz: Musikverlag Schott (and other publishers), 1992.

Himmelfarb, Gertrude. *One Nation: Two Cultures: A Searching Examination of American Society in the Aftermath of Our Cultural Revolution*, 2nd ed. New York: Vintage, 2001.

Jerger, Wilhelm. *Die Wiener Philharmoniker, Erbe und Sendung.* Vienna: Wiener Verlag, 1942.

Kerner, Leighton. »Prosit Neu Jahr Mit Firecrackers.« *Village Voice* 30 (22 January 1985): 97.

Kralik, Heinrich. *The Great Orchestra: The Vienna Philharmonic and Their Conductors.* Vienna: Frick, 1955.

Lamb-Faffelberger, Margarete. »Beyond ›The Sound of Music‹: The Quest for Cultural Identity in Modern Austria.« *German Quarterly* 76/3 (Summer 2003): 289-99.

Orwell, George. *The Collected Essays, Journalism & Letters of George Orwell*, vol. 4, »In Front of Your Nose, 1945-1950.« Eds. Sonia Orwell and Ian Angus. Boston: Nonpareil, 1968.

Osborne, William. »Ozawa Conducts the Vienna Philharmonic's New Year's Concert: Tokenism and Public Relations.« *IAWM Journal* 8/1-2 (2002): 21-3.

Potter, Tully. »CD Reviews – Historic New Year's Day Concerts.« *International Classical Record Collector* 5 (Summer 1999): 81-2.

Saffle, Michael. »›Do You Ever Dream of Vienna?‹ America's Glorification of Musical Central Europe, 1865-1965.« *Identität • Kultur • Raum: Kulturelle Praktiken und die Ausbildung von Imagined Communities in Nordamerika und Zentraleuropa.* Eds. Susan Ingram, Markus Reisenleitner, and Cornelia Szabó-Knotik. Vienna: Turia + Kant, 2001. 59-76.

_____. »From Schlagobers to Schoenberg: Cultural Transfer, Identity and Otherness, and Images of Musical Vienna in the New York Times, 1918-1939.« *Reverberations: Representations of Modernity, Tradition and Cultural Value in-between Central Europe and North America.* Eds. Ingram, Reisenleitner, and Szabó-Knotik. Frankfurt and New York: Peter Lang, 2002. 63-89.

_____. »Traveling Tales: The Trapp Family Singers and Maria von Trapp's Post-World War II Immigration Narratives.« *Ports of Call: Central European and North American Culture/s in Motion.* Eds. Ingram, Reisenleitner, and Szabó-Knotik. Frankfurt and New York: Peter Lang, 2003. 67-100.

Stevens, Wallace. *Poems*, ed. Samuel French Morse. New York: Vintage, 1959.

Strasser, Otto. »Wiederaufbau und Konsolidierung.« »Den Wiener Philharmonikern zum Geburtstag.« *Österreichische Musikzeitschrift* 22/2-3 (February-March 1967): 112-28.

Szabó-Knotik, Cornelia. »Musikalische Elite in Wien um 1900: Praktiken, Prägungen und Repräsentationen.« *Identität • Kultur • Raum: Kulturelle Praktiken und die Ausbildung von Imagined Communities in Nordamerika und Zentraleuropa.* Eds. Ingram, Reisenleitner, and Szabó-Knotik. Vienna: Turia + Kant, 2001. 41-58.

The New York Times, 1986.

Tosic, L. »Neujahrsswing im Schanitakt: das Vienna Art Orchestra macht den Philharmonikern Konkurrenz – Am. 1.1.2000 findet ein alternatives Neujahrskonzert mit verjazzten Strauß-Stücken statt.« *Bühne* 12/12 (December 1999): 50-1.

Weigel, Hans. *Das Buch der Wiener Philharmoniker* (with accompanying phonorecording). Salzburg: Residenz, 1967.

»An Unruly Completeness«:

Fritz Lehner's *Mit meinen heißen Tränen*[1]

JOHN C. TIBBETTS

»Films about Franz Schubert have presented only a clichéd, unrealistic portrait of the composer,« says film director Fritz Lehner in a recent interview.[2] »I grew up with that, and I don't like it. When I made *Mit meinen heißen Tränen*, I wanted to show Schubert as a human being, not as some sort of monument, god-like, untouchable. He had been a teacher, and I also had been a teacher; and I was drawn to that.«

Mit meinen heißen Tränen (1986) provides a case study of how the film medium can deploy its audio-visual resources to both capture and dramatize biography and history in ways that resist and revise the preexisting stereotypes of print and visual discourse. It rechannels the lineaments of a hitherto cozily familiar physical and cultural milieu of Schubert's Biedermeier Vienna – a term borrowed from an 1855 volume, *Poems of Gottfried Biedermaier*[3] – into a series of images and sounds that, in the words of historian Simon Schama, emphasizes the »obstinate unfamiliarity« and »otherness of the past« (41). Schubert and his Vienna differ markedly from the conventional image, prevailing since Schubert's day, of an amiable little man in a cheerfully tidy Vienna who tosses off musical masterpieces in coffeehouses and salons while sighing in vain for the love of his lady fair. »*Mit meinen heißen Tränen* introduced a darker Schubert,« declares Schubert biographer Christopher Gibbs, »– not simply suffering, but alienated, ill, and isolated even among family and friends. It created a considerable stir in Europe« (54).

At first glance, Schubert's life and circumstances are hardly prepossessing. Born in Vienna in 1797, the twelfth child of a humble schoolmaster, he sang as a boy in the Imperial Chapel's chorister, and, after a desultory attempt at teaching in his father's school, pursued a composing career. Although he was encouraged by a small but enthusiastic circle of artists and writers, many of his major works – including operas, symphonies, religious music, and chamber works – failed to find a wide public and remained unpublished at his death in 1828 at age 31. Only his songs and some of his dance music were generally known among his contemporaries. Indeed, he held only one public concert of his works, in 1828. He traveled seldom, never held an official appointment, never married, earned only enough to survive, and died in 1828 of typhoid and complications of tertiary syphilis.

Yet, even before his death Schubert's life and circumstances were wrenched from his private space and thrust into the more public arena of collective myth. Cultural tastemakers from Weimar to Hollywood have busily constructed their own versions of his life and milieu, establishing him, in the words of historian Robert Waissenberger, as »the prototype of the Biedermeier artist« (254). »No composer, not even Mozart, has been more often reinvented and reinterpreted,« writes critic Peter G. Davis, »or given so many different faces« (59). The relative absence of biographical data is mostly at fault. Many contemporary accounts, memoirs and anecdotes from friends and colleagues assured us he was a cherubic idiot savant – »a guileless child romping among giants,« as Robert Schumann famously observed – a modest, relatively untutored *Schwammerl* (»little mushroom«), who sat in coffeehouses and effortlessly improvised dance tunes for his amiable circle of friends, but who was unlucky in life and love and who died young, alone and neglected. This fiction was first institutionalized in 1865 by his first biographer, Heinrich Kreissle von Hellborn, a work which was embellished and translated into English by Edward Wilberforce a few years later; and it was given its fullest elaboration by George Grove in 1882.[4] As late as 1949, biographer Robert Haven Schauffler was still painting this cozy portrait of an »Ariel« of music, happily singing his songs amidst his loving circle of friends: »Gathered about a poor, ugly, insignificant-looking little school assistant of proletarian birth, a creature without physical charm or sexual attraction, was a circle of brilliant and absolutely devoted young men, who found their greatest pleasure in supporting and encouraging him, and furthering his genius in every possible way« (74).

Shortly after the turn of the twentieth century began a flood of fanciful novels, operettas, and films further embellishing the Schubert myth. First appeared Rudolf Bartsch's novel, *Schwammerl* (1912), which was adapted by Alfred Willner and Heinz Reichert into a *singspiel* in 1916, *Das Dreimäderlhaus* (*The House of Three Maidens*), with melodies adapted by H. Berté; and in turn transformed in 1923 into a popular London musical show, retitled *Lilac Time*. In Paris the show, somewhat modified, was known as *Chanson d'Amour* and in Italy as *La casa delle tre ragazze*. And in America it was further altered and again retitled in 1921 by Sigmund Romberg and Dorothy Donnelly as *Blossom Time* (the generic title I will use throughout this discussion). The basic storyline, a variant of *Cyrano de Bergerac*, remains the same: Schubert is a penniless composer whose love for a young singer (named, alternately, Hannerl, Mitzi, or Vickie, etc.) is frustrated in his affections by a rival, the handsome aristocrat, Baron Schober. Ironically, when the Baron sings her the love songs Schubert has composed for her, she falls in love with the Baron, not Schubert. The heartbroken Schubert sacrifices his own feelings and plays cupid to the lovers. He then retires alone to his garret where, according to any of the particular versions cited above, he either fails to complete his Eighth Symphony or he composes one last song, »Ave Maria,« before he expires.[5]

Blossom Time, in its various incarnations, has been a perennial stage hit with many Schubert melodies garnering »pop« song status with their new lyrics. »Critics almost unanimously praised Romberg's skillful adaptation of Schubert's classical compositions to the requirements of the popular musical,« writes Broadway musical historian Abe Laufe. »Such songs as ›Serenade,‹ ›Ave Maria,‹ and ›Song of Love‹ – which became the hit of the show – helped make *Blossom Time* one of Romberg's most successful operettas« (Laufe 12-13).[6]

After the appearance of a silent film version of *Das Dreimäderlhaus* in 1917, directed by Richard Oswald, a host of sound films brought variants of *Blossom Time* to the talking pictures. These included an Austrian film, *»Leise flehen meine Lieder«* (1933), starring Hans Jaray as Schubert and Marta Eggerth as his love interest, Caroline Esterhazy; the British-made *April Romance* (1936), a vehicle for the ear-splitting voice of opera star Richard Tauber in his signature role and Jane Baxter as the beautiful Vickie; a French production, *Serenade* (1939), with

Bernard Lancret as the composer and Lillian Harvey as an English dancer; the joint British-American *Melody Master* (1941), starring Alan Curtis and Ilona Massey; a French version, *La belle meunière* (1948), with Tino Rossi and Jacqueline Pagnol; an Austrian production, *Franz Schubert: Ein Leben in zwei Sätzen* (1953), starring Heinrich Schweiger and Aglaja Schmidt; and another Austrian film, *Das Dreimäderlhaus* (1958), with Karl-Heinz Boehm and Johanna Matz. Shot on clean, well-scrubbed sound stages and on location at a handful of picturesque exteriors, they all display, like their stage counterparts, the familiar, sanitized, cozy milieu of the Biedermeier stereotype. Schubert himself is manly enough, trim, handsome, healthy, and markedly heterosexual. When he's not marching around the woods bellowing »Das Wandern« with his students, as in *Blossom Time*, he and his pals, Mayrhofer, Vogl, von Schwind, and Schober roister about in the taverns, bustle about their kitchen frying chickens, dance on the village green, and serenade the apple-cheeked girls (sometimes accompanied by a convenient backup chorus) with a medley of Schubert's tunes in *Das Dreimäderlhaus*. Schubert is rarely seen actually working at composing. »How do you get your ideas,« he is asked in *Das Dreimäderlhaus*. »I don't know,« he replies humbly. »They just come from a feeling deep inside.« Stoically, he shrugs off the indifferences of publishers. »My music may be good or bad,« he says in *Franz Schubert. Ein Leben in zwei Sätzen*. »But it is what it is.« And when his lady fair turns away from him in favor of his best friend, Schubert murmurs resignedly, in *Melody Master*, »A man must turn his face to God« and suddenly expires at his writing desk, the strains of »Ave Maria« welling up from an invisible orchestra and chorus.[7]

So numerous are these stage, screen, and television biopic versions of the *Blossom Time* model that Schubert might possibly qualify as the most frequently dramatized of all composers, both classical and popular, possibly excepting Franz Liszt[8] and Wolfgang Amadeus Mozart. And, of course, not the least of his attractions is a readymade storehouse of memorable melodies available to film composers like Miklos Rozsa and Tin Pan Alley tunesmiths like Sigmund Romberg. Fragmented and fetishized, the tunes are transformed literally into what historian Christopher Gibbs calls »the soundtrack for mythological narratives of his life.« To this day, Gibbs continues, we are assured that »even with so many miserable circumstances, Schubert's music laughs through its tears and the maudlin conflation of his life and works in myriad biographies and fictional treatments makes [audiences] past and present weep« (36).

Revisionist historians, including Maynard Solomon, Joseph Horowitz, and Leon Botstein; Schubertian performing specialists, such as Alfred Brendel and the duo-piano team of John and Richard Contiguglia; and events such as the series of *Schubertiaden*, held in New York City from 1988-1997, point to a reality significantly different from the »Blossom Time« paradigm. »You cannot identify another famous composer whose popular portraiture is so [...] wide of the mark,« opines Joseph Horowitz, director of the New York *Schubertiaden* and dedicated to revising and correcting prevailing Schubertian images.[9] Hardly a composer of limited academic learning and compositional ignorance, Schubert was, according to the Contiguglias – who have devoted more than forty years to championing Schubert's four-hand piano literature – »a composer of extremely sophisticated counterpoint,« an »intellectual of sorts« who »gathered with friends to do more than just make music, but also to read aloud poetry and discuss the philosophical currents of the day in Germany and Austria.«[10] However, because of the movies' cozy and tender mythology, »people have become so possessed by this visual picture that they have lost sight of the complexities of the man and the darker implications of much of the music.«[11] Images of a cherubic innocent, free of guile, perpetually frustrated in his heterosexual relationships, are countered by Schubert's own confessional letter of 31 March 1824, to his friend, Kupelwieser, in which he movingly confessed, »Imagine a man, I say, whose most brilliant hopes have perished, to whom the felicity of love and friendship have nothing to offer but pain at best, whom enthusiasm [...] for all things beautiful threatens to forsake [...]« (quoted in Solomon 206). In his groundbreaking, albeit controversial study, »Franz Schubert and the Peacocks of Benvenuto Cellini,« Maynard Solomon references the gay subculture in which Schubert moved, his apparent lack of interest in women, his own ambivalent sexuality, his moods of alienation and suicide expressed in his letters, egomaniacal outbursts among his friends, his relative success in publishing much of his music, the innovative/problematic nature of his last experimental compositions, and his contraction of a fatal syphilis. After noting the prevalence of prostitution in the Vienna of Schubert's day, and after exhaustively reviewing testimonies to this effect by contemporaries, Solomon declares, »It is worth exploring the possibility that Schubert's friends were suggesting not simply that he was sexually promiscuous, but that his promiscuity was of an unorthodox character« (194). Chief among those male friends who may have pursued a homosexual relationship with the composer was Franz von Schober (a character, we recall, portrayed very differently as Schubert's rival for the ladies in the »Blossom Time« paradigm). »I believe,« furthers Solomon, regarding Schober, Mayrhofer, Schwind, and others, »it is reasonably probable that their primary sexual orientation was a homosexual one« (202). Their behavior may have contributed to the marginalization and alienation Schubert felt so deeply: »Schubert and his companions inhabited a clandestine realm, one constantly beset by a variety of fears – of surveillance, of arrest and persecution, of stigmatization and exile« (197).[12]

Underlying it all is a portrait of Vienna quite at odds with the cozy, domestic milieu depicted on stage and in the movies. In his portrait of Schubert's last years, Alfred Brendel reminds us that the Viennese musical establishment largely ignored the composer, during and after his lifetime. Brendel quotes Schubert's own less than enthusiastic opinion of his native city in a letter dated 1827: »It is admittedly rather big, but makes up for it by being devoid of cordiality, openness, true thought, sensible words, and, in particular, spirited deeds« (quoted in Brendel 32). The city itself, writes historian Leon Botstein, which has long been in the grip of a nostalgia for the pre-1850s – »before industrialization and modernization, bankers, businessmen, Jews and Czechs« – was a nasty, blunting, and dangerous place to live: »The nasty and alluring details of the Vienna of the 1820s can go a long way in illuminating complex and hidden dimensions of Schubert's work« (19). Botstein notes that in Schubert's time, under the strict control of Prince Metternich, the city had grown to a population of 300,000, and the density of occupation averaged 38 persons per household. This was an era of state censorship of all intellectual and political activities, declining living standards, deteriorating sanitary conditions, overcrowding, dire poverty, and public begging. Garbage spilled into the streets, sewage flowed freely, refugees and beggars from the late Napoleonic wars crowded the streets, and houses were like honeycombs of narrow, crowded rooms. Cholera was in the air. »Death and its ceremonial rituals – requiems, funeral processions, and burials – were a visible and regular part of daily life,« continues Botstein. »Schubert's death at an early age was itself not extraordinary« (28).

This contentious attitude toward the hitherto accepted hagiographical formulas of the *Blossom Time* paradigm is typical of a general revisioning of the numerous biopics of composers, politicians, athletes, inventors, soldiers and statesmen so prevalent in Hollywood and Europe during the so-called »classical studio« years of the 1930s through the mid-1970s. It can be detected in the change in attitude, tone, and graphic detail in dramatized composer biographies appearing in the post-1970s era.[13] The breakup of the classical studio oligopoly, post-Vietnam disillusionment and cynicism, collapsing political regimes, reconfigured nation-states, postmodern interrogations of fact and fiction, and global disasters like terrorism and the AIDs epidemic revised conventional attitudes toward history and biography. No longer »contained« by stifling story formulas and censorial restrictions, the composer subjects burst their stereotyped frames, climbed off the mountain top, and, like Frankenstein's monster, revealed their scars and ran amok. British filmmakers Ken Russell and Tony Palmer led the way, replacing the canon of sentimental, love-struck, romantically mad and tragically failed artists with a gallery of fools, cads, opportunists, failed dreamers, and societal victims.[14] The Biedermeier-like worlds of countless biopics about Franz Schubert and Johann Strauss, Jr. collapsed, revealing undercurrents of political repression, spiritual corruption, creative failure, and nihilistic despair.

Mit meinen heißen Tränen is only one of many examples. A few more that appeared prior to it include Ken Russell's *Dance of the Seven Veils* and *The Music Lovers* (both 1970), bitterly sardonic explorations of the darker political and sexual aspects informing the life and music of, respectively, Richard Strauss (Christopher Gable) and Pyotr Tchaikovsky (Richard Chamberlain); Renato Castellani's *Life of Verdi* (1982), a postmodern interlacing of the narrative conventions of traditional historiography, performance, and multi-vocal testimony concerning the career of Giuseppe Verdi (Ronald Pickup); Milos Forman's *Amadeus* (1984), an experiment in the psychological aspects of memory (Antonio Salieri portrayed by F. Murray Abraham) and identity (Mozart, portrayed by Tom Hulce); Peter Schamoni's *Spring Symphony* (1985), a dual portrait of the dysfunctional marriage of Robert Schumann (Herbert Groenemeyer) and Clara Wieck (Nastassja Kinski); and Tony Palmer's *Testimony* (1987), a dissection of the symbiotic relationship between Dmitri Shostakovich (Ben Kingsley) and Josef Stalin (Terence Rigby).

Fritz Lehner's *Mit meinen heißen Tränen* (*With My Hot Tears*), first seen on Austrian television (ORF) in the fall of 1986, occupies a unique position, not just among the many Schubert biopics, but in the composer biopic genre in general.[15] It stands out for the complexity of its psychological portrait of an artist and in the meticulous detail of its visual and aural depiction of the Viennese milieu. Shot over a period of two years at an estimated cost of four million US dollars, it was broadcast on three successive days on Austrian ORF television, from 31 October to 2 November 1986. It was also broadcast or shown theatrically in 26 countries, including the United States, where it was presented at Lincoln Center in an abridged version, *Notturno*, in 1992. A year later, it was featured at the 1993 *Schubertiade* in New York City.[16] Nominated for the *Prix Européen du cinema* in 1988 and winner of a dozen prizes, including the *Camera d'Or* at the *Festival de Cinema de Barcelona*, it both amazed and outraged its viewers and critics, as will be seen.

The project had a long gestation. »I began thinking about a Schubert film as early as 1983,« Lehner says. »I studied his music, looked at 2,000 paintings and researched dozens of books about him and his period, particularly the work of Otto Erich Deutsch. But I didn't want to make a documentary or an educational film. I wanted to make a dramatic story. You know, Deutsch once said he could account for only *five weeks* out of Schubert's entire life. Everything else is either unknown or open to speculation.«[17] As Theophil Antonicek put it, in his afterword to the published screenplay of *Mit meinen heißen Tränen:*

> Fritz Lehner's film endeavors to illuminate, to break open the hidden realm of »Schwammerl's« [i.e., Schubert's] final years. […] But Lehner's film isn't about documentary sources; rather, it employs the methodology of film-making in order to awaken the viewing public from its familiar and lulling impressions of Schubert. […] It

Fritz Lehner is a native of Austria and works as a free-lance film director.

goes without saying that Lehner has mastered and drawn upon the available factual information [about Schubert's life]. But the facts are little more than a jumping-off place or a set of stimuli for heartfelt artist's fantasy.[18]

Thus, Lehner's film, like its predecessors, has not stinted on its speculations about incidents in Schubert's life; but unlike them, it casts a colder, sharper, more knowing gaze on its subject. »The great majority of paintings and films about the period portray the Biedermeier ideal of Schubert and his Vienna,« he says, »and not the harsh realities, the poverty, the disease, the unsanitary conditions, etc. In that time, the censorship of Metternich refused to acknowledge publicly those conditions. Yes, many of the scenes in the film are my own invention, but they were always based on *what could have happened*. And I don't think anyone can contradict what I have shown, or prove that it could *not* have happened!«

A brief synopsis is in order. *Mit meinen heißen Tränen* consists of three parts: »*Der Wanderer*« (»The Wanderer«), »*Im Reich des Gartens*« (»In Gardenland"), and »*Winterreise*« (»Winter Journey«). »*Der Wanderer*« opens in the summer of 1823 (*Tränen* 9-47). Schubert (Udo Samel) is in a hospital being treated for second-stage syphilis. A blurred image coalesces into a sharply focused close-up of Schubert's disease-ravaged head. The hair is falling out in tufts. He is rebuked by a visiting monk for the sin of promiscuity. Upon leaving he dons a wig, which subsequently becomes an object of jest among his friends. At home, he speaks with his stepmother, who bestows upon him a prolonged kiss that is decidedly erotic and incestuous in its implication. Outside, the region of the *Blutgasse* is teaming with carriages, soldiers, citizens, and beggars. Sewage and garbage spill out of the gutters. Schubert sits by the curb a moment, a motionless figure amidst the bustle and noise, listening to the bells of St. Stephens. He recognizes a young prostitute and runs to confront her. In a silent rebuke, he lifts his wig to reveal the sores on his scalp. »I'm sorry,« she says after a moment and walks away.[19] All the while, a legless man dressed in rags and trundling about in a little wheeled cart, is following him…

A few hours later, Schubert finds himself with a party of friends traveling by coach to a party at Mauritz von Schwind's country home. Following behind the coach is the legless beggar. Although self-conscious about his new wig and stinging from an occasional sly insult, Schubert entertains them by playing some of his German Dances at the piano. At his feet sits the beggar. All the while he is envious of his friend and roommate, Franz von Schober (Daniel Olbrychski), who is flirting with a female companion. When Schober tries to take advantage of the woman, who is obviously drunk, Schubert angrily confronts him. He kicks Schober viciously, and the two grapple in a rough tussle. »Take revenge on your prostitute, not on me,« Schober says afterward. »Or are you jealous? Be glad that I took you with me to Spittelberg; otherwise you might still be a virgin [...] Now, you leave my rooms immediately.«[20]

Schubert returns to town, alone, and goes to his father's school. »I need your help,« he says ruefully. But Father Schubert only gestures with his stick and knocks askew his son's wig. The children in the classroom laugh scornfully. »He was once my son and a teacher, like me,« his father tells the students, »but now he keeps company with filthy whores« – he whacks Schubert with his cane. »Nineteen children have been born to me; ten have been taken away. And you are left alive.«[21]

Night grows near. Utterly crestfallen, Schubert crouches at the graveside of his mother. He lifts his face to the twilight sky and removes his wig. »I'm talking to you, mother; do you hear me?« he soliloquizes softly. »You rot in your grave and I am rotting alive. You got uglier with every child, and I inherited the ugliness. Why didn't I die from that? I watch my fingers like dough on the piano. Young girls close their eyes against me. My beautiful music – it's just markings on paper.«

Back at Schober's apartment, Schubert prepares to leave. But as he packs his trunk, a repentant Schober renews his pledge of friendship and invites him to stay. Wordlessly, Schober takes out a small knife, pricks Schubert's finger to draw blood, spills drops of the infected blood into a wine glass and then, deliberately, meaningfully, drinks it down.

The second part, »*Im Reich des Gartens,*« transpires in one day in autumn 1827 (*Tränen* 51-89). Schubert accompanies his friends, Mayrhofer, Schober, von Schwind, Hüttenbrenner, and others on an outing in the country. The main action concerns Schubert's brief but intense infatuation with a young woman,

Magdalena (Therese Affolter), who, he learns, has been hired by his friends to be his companion. Incidents include a wild coach ride through the forest, a fleetingly romantic interlude with Magdalena on a grassy meadow, an excursion to a waterfall, a trip downriver by boat, a visit to an empty church, and a penultimate night scene of a band concert by Johann Strauss, Sr.

Brief but telling vignettes suggest the darker undercurrents beneath the general air of loose hilarity. For example, for the first time there is a hint of homosexual behavior, when Schubert rebuffs overtures by his friend, Hüttenbrenner. During their walk through the forest, they watch as a runaway gypsy is hunted down by mounted Austrian troops and shot dead. Meanwhile, a surrealist touch is added when a hot-air balloon, to which is tethered a goat, passes by overhead. The waterfall sequence is haunting in its ambiguity: when one of the group breaks a local rule and plunges into the cataract, the waterfall suddenly, mysteriously, stops its flow. As if in a trance, Schubert ventures into the rocky cliff through a cleft in the rock. He stands alone in the dimly lit grotto, surrounded by muffled and indistinct sounds. When he reemerges, the rushing torrent of water resumes.

Most arresting of all is the night scene when, amidst the concussions and showers of skyrockets, Schubert clumsily tries to make love to Magdalena. When she shrinks from him, he realizes that Schober has paid her to be his companion that evening. »It's better to let you know,« she confesses. »I don't get that much money every day. If you want, I'll lie down with you all night.« He staggers back from her. »We all love you,« she repeats as he breaks into a run.[22] His retreat blocked by the river, he stands knee-deep in the water, the fitful light of the fireworks limning the abject distress on his face. »I am Franz Schubert,« he cries out to the indifferent sky and the curious onlookers crowding the riverbank. »I AM FRANZ SCHUBERT!« he cries again in a mounting frenzy.[23] And then again… By contrast, the quiet epilogue reveals Schubert, alone at break of day, wandering through the party debris scattered across the grounds. Slowly, deliberately, the camera tracks across the meadow, revealing a random assortment of overturned chairs, an abandoned mirror, a tethered goat, a clutch of von Schwind's erotic sketches, a snake trapped inside a bottle, etc. The party has ended, not on a note of hilarity but with an elegy.

Schubert joins his friends on an outing in the second part of Mit meinen heissen Tränen, *»Im Reich des Gartens.«*

The concluding third of the trilogy, »*Winterreise*,« transpires over a period of a few weeks in November 1828 (*Tränen* 93-132). The action begins as Schubert moves into rooms adjoining the apartment of his brother, Ferdinand (Wolfgang Hübsch). A preternatural quiet envelops the action like a shroud. Powerful close-ups intensify a sense of mounting claustrophobia. The cramped, box-like interiors seem to shrink in dimension. The piano room narrows down to the nook enclosing Schubert's tiny writing desk; which is succeeded in turn by the even smaller bedroom, where his fatal convulsions are restrained by attendants and a priest.

The first real intimations of disaster come in a harrowing scene, the more remarkable for its seeming stasis and calm. After picking out the opening measures of the Andante of the A major Piano Sonata, Schubert leaves the piano and moves to his desk by the window, where in utter and absorbing silence he bends over the manuscript, rules the staff lines, and picks up his quill pen. Slowly, gradually, a montage of extreme close-ups and selective sounds: the rattle of the windowpane, the flutter of a moth at the candle flame, the scratch of the quill, a fly tracking ink across the paper, a shot of another fly trapped in the inkwell – isolate and encapsulate the moment. Then, as we share Schubert's internal diegesis, i.e. the sounds of the music itself – the slow build-up to the impassioned outburst midway through the Sonata's Andante – are heard on the soundtrack. He stops and reaches for another piece of paper. He inscribes the treble clef on the manuscript. But the design is not right. He pauses uncertainly. He tries to draw it again. No. Has he forgotten how to draw it? He's distracted now, puzzled. The ink splatters. He searches frantically through a drawer for another pen. Suddenly, abruptly, he starts back from the desk, overturning the chair. The camera retreats from him, framing him in a long shot, writhing in anguish, the lowering, shadow-splashed ceiling dominating the frame and diminishing his figure. The music grows in intensity. He staggers, then collapses to the floor, the strains of several stark chords hammering away on the soundtrack…

Schubert is in trouble. But life goes on, uncaring, indifferent, within and without the tiny apartment. Josefa (Michaela Widhalm), Schubert's 13 year-old half-sister, is always in the room with him, distracting him as she hums snatches of his melodies and plays with the cat. Through his windows he catches glimpses of the world below and beyond his reach – a clockmaker, a woman tending a flowerpot, birds on a roof, a man and a baby, a little boy getting a haircut, a bird in a cage. A figure stands below his window and returns Schubert's gaze. The face is Schubert's own – a ghostly *Doppelgänger*. Later, friends come to visit. One of them stumbles through the B-flat Sonata's first movement. »Don't you have anything easier?« he asks. »With music like that you won't be successful. Look at Paganini, Strauss, Rossini, they get more money from one concert than you get in a year.«[24] Schubert isn't listening. He sings a verse of »*Die Krähe.*« Ferdinand enters and closes all the windows. »He burns my clothes,« says Schubert. »He's scared that I would infect him and his wife. I'm glad that you came, Spaun – or didn't you know? Best get away from me. It's even worse that you dislike my music.«[25]

Brother Ferdinand massages his shoulders.

»Watch it, don't get infected,« snarls Schubert. He leans down to the keyboard and plays the opening chords to »*Der Leiermann.*« He murmurs to Ferdinand in an aside, »Now go away. You can burn my clothes if you wish. As if dying would be the worst thing. I'm just sad because I don't know what's really wrong with me. Is it my old illness? Nervous fever? That's how my mother died.«[26]

Later that night. Schubert gazes across the street at a woman framed in her apartment window. Unflinchingly, he returns her gaze as she slowly disrobes, her body outlined behind the transparent nightdress. He scribbles drawings onto the pages of a book. The next morning the woman comes to his room. »I was curious what you see when you look my direction,« she says, gazing out the window back to her apartment. »I also watch you writing. Are you a poet? Do you write music to dance to?«[27] She opens his copy of Heine's *Buch der Lieder* and looks dispassionately upon the series of obscene drawings he's scrawled on the page. She pauses, not a flicker of a reaction. »Will you dance with me?« she asks at length. »Come on. I know you want me.«[28] They dance, his steps hesitant at first, then more vigorous. They stop, breathing heavily. »We'll dance again soon, I hope,« he says. She replies, »Just wave; you watch me all the time anyway.«[29]

In bed, Schubert is writing a letter. »Dear Schober,« it reads. »I've not eaten in eleven days. Come and help me out with some reading, Cooper's *Last of the Mohicans* and *The Spy*.«[30] He looks up. Josefa is hanging up clothes and putting wood onto the fire. He watches her with growing interest. She comes to him and bathes his forehead. He strokes her hair. Then, almost casually, he kisses her forehead, then her lips. She lies down beside him. Music from the second movement of the Ninth Symphony wells up on the soundtrack as they embrace in a frantic and clumsy frenzy. After a few moments, they pause, his face contorted, her breathing heavy.

Later, alone, he discovers his brother has locked the door against the other apartment. Schubert is trapped, entombed. He peers out the window and sees his own figure on the street below. He shrinks back. He takes off his coat and sits down on the bed.

The doctor is at his bedside measuring out something. »We have a new physician, Franz, and he says that you'll get better again,« says Ferdinand.[31] Later. An incoherent Schubert is singing as a priest arrives. »Ferdinand, if you're still my brother, send him away,« he retorts. »If I never needed him before, I don't need him now. Ferdinand, I always liked you most, but you have to promise not to prepare me like they did Beethoven, when they cut off his hair, or cut off his head, like Haydn.«[32] Ferdinand answers, »Don't worry, the pipe will taste good again. You're getting better every day. You sang so beautifully just now.«[33]

Josefa, hearing his delirious voice from the street, quickens her steps, sloshing the wine, spilling it as she runs up the stairs. Inside, she finds him singing while being bathed by an attendant. His mouth is chapped and blistered. Convulsions set in and she retreats to the outer room, where the priest waits. Schubert continues to sing deliriously. The others rush in and close the door, blocking her last view of him being restrained by the priest. She is pushed away and the door closes.

I have indulged in a certain amount of detail in the foregoing synopsis because, although it hardly scratches the surface of this bleak, meticulously detailed and dramatically complex portrait of Schubert, it perhaps is enough to convey its superbly textured aural and visual qualities and to help us understand the outrage it has elicited in some quarters. »There were many controversies and protests about the film,« confirms Lehner. »People were angry that I had ruined the clichés of *Das Dreimäderlhaus*. People didn't want to admit that Schubert had sexual desires, that he had contracted syphilis, that he had to wear a wig, that he felt alienated from his friends, etc. I even received some death threats!«

Udo Samel's portrayal of Schubert is extraordinary in its nuanced and quiet subtlety. His physical resemblance to extant portraits is uncanny. »I looked for a year for just the right person to portray Schubert,« says Lehner. »Udo was an actor at the *Schaubühne Theater* in Berlin, working under the directorship of Peter Stein. He had done only one film before coming to my Schubert project. He had an astonishing resemblance to Schubert and is an excellent actor. He's also a musician who plays the piano and once had ambitions to be a composer. I had to postpone the shooting for a year in order to get him.«

Clearly, this is a far cry from the sentimentalized, sweetly suffering Schubert of the *Dreimäderlhaus* paradigm. From the very beginning of the first episode, he is depicted as a marginalized figure, diseased, lonely, existing at the fringe of a circle of young men and women. His ill-fitting wig literally identifies him as

Before coming to the role of Franz Schubert, Udo Samel was an actor at the Schaubühne Theater in Berlin.

Courtesy of Gabriela Brandenstein.

a »misfit« everywhere he goes. Music provides him only limited solace. Scenes of him at the piano are few, and the music is either tolerated by his friends as amusing entertainment or largely ignored altogether. Episodes in the third part confirm that his newly composed music finds neither favor nor support among his circle. He is hopelessly estranged from his family. His mother is dead, and his father mocks him. His emotional life is stunted. »Normal« love, as well as the free-and-easy promiscuity of Schober – the only companion whose presence is strongly marked in all three parts and who has obviously drawn him in the past into illicit sexual encounters – is denied him. His only sexual activity here is limited to the ill-fated involvement with a prostitute in the first part, the aborted romance with Magdalena in the second, and the voyeuristic encounters with the woman in the window in the third. Otherwise, there are only the incestuous implications of the prolonged, intimate kiss exchanged with his mother-in-law in the first part and the frenzied grapplings with his sister-in-law in the third part. The issue of syphilis, long a controversial topic in Schubert studies, is announced at the very beginning and becomes a metaphor for the AIDS epidemic of today. In this light the penultimate scene in the first part, when Schober renews his pledge of friendship with the blood-infected glass of wine, takes on a particularly tragic significance for both of them. Thus, harbingers of death are omnipresent – from the presence of the legless beggar who follows him throughout the first part to the confrontations with his ghostly self in the *Doppelgänger* scenes in the third.

Lehner defends his inclusion of such scenes. »It is true that Schubert's father disliked his friends and was ashamed of Franz's illness,« explains Lehner. »Based upon that kind of evidence, I showed Franz' father insulting him and turning his back on him. The character of the beggar was drawn from the many homeless, crippled men who returned from the Napoleonic wars to form little conclaves just outside of Vienna and who came into town to beg. I wanted to show the poverty that people like that endured, and how they had to survive on the trash and food thrown into the streets. I knew little about AIDS at the time I made the film, but I can understand why people might infer that from the scene with Schober and the infected wine glass. I wanted to suggest a homoerotic implication to their relationship – there are hints of that everywhere – but not necessarily that Schubert was actively homosexual. It was just a time when men displayed affection for each other more openly than today. What I really wanted to show was that Schober's gesture with the wine glass demonstrates that he feels responsible for Schubert's illness – it was he who took Schubert to the whore – and that now he shares with him the certainty of death.[34] It's all part of the romantic sensibility of the time: we share life, we share death. You see the same thing in Schober's relationship with the young girl at the outdoor party, who has infected lungs. He kisses her, knowing full well that he might become infected.«

The most controversial scene, perhaps, the erotic encounter with Josefa, is likewise defended. »Many people have asked me about that scene. It is documented that she really took care of him in a very loving way in the last weeks of his life.[35] I wanted their scene together to suggest innocent love, pure love. But it is a love that will remain unfulfilled. I suggested the same thing in the scene with the woman in the window. When she comes to him, he is unable to dance with her. It has been documented that he *couldn't* dance. All of which is sadly ironic, since he wrote so much about love and about dance.«

Other scenes preserve their mystery, and Lehner is content with that. »The scene at the waterfall is another invented scene. People really believed in miracles at that time. The fact that the waterfall ceased to flow could have a perfectly normal explanation – probably that work of laborers upstream caused it to happen. But Schubert's friends preferred to regard it as a miracle, maybe somehow connected with Schubert's innocence, or with the flow of his creativity. And the strange sounds Schubert heard inside the grotto were probably only the winds acting like a wind harp, although you might want to hear them as musical ›sounds‹ in his head. All I wanted to do was create a scene open to interpretation.«

Meanwhile, we are left with myriads of tiny moments, images, and sounds that come and go, fleeting, ineffable, inexplicable – the goat tethered to the hot-air balloon, the snake winding its way through the neck of the bottle, the runaway gypsy, Schubert's crudely inscribed obscene drawings, the visit to the empty church, the violent fight with Schober, the whisperings in the grotto, and so many others too numerous to cite here. Lehner's film belongs to a select company of historical films that reconstructs a vanished world which displays what historian Simon Schama describes as presenting »an unruly completeness.« Schama applauds this respect for »the strangeness of the past.« These are films, he continues, »that embrace history for its power to complicate, rather than clarify, and warn the time traveler that he is entering a place where he may well lose the thread rather than get the gist« (41). If moments in *Mit meinen heißen Tränen* pass by enigmatic and unexplained, the sense of the strangeness of the past is all the richer.

Indeed, Lehner's meticulously realized portrait of Schubert's environs contributes more to its mystery than to its more familiar Biedermeier images. This unvarnished depiction of Viennese life is grim and graphic. »I used many locations in Vienna and in the Vienna Woods and in the Donau Auen meadows next to the Danube, about thirty kilometers outside Vienna,« explains Lehner. »We do know that Schubert actually went to these locations. Some of the first part I shot in the first district around the *Alte Universität*, the street called

Bäckerstraße and the *Schönlaterngasse*. Most of the third part was shot in the studio, except for the street scenes outside. The house opposite Schubert's room was also shot in the studio. I copied in the studio the room where Schubert died.«

But what is most astonishing about *Mit meinen heißen Tränen* is, paradoxically enough, its *silence*. It is undoubtedly the quietest movie about a composer ever made. Just as silence can »speak« eloquently in music, so here it commands our attention in the film. Granted, there are many scenes featuring music, all of them appropriately chosen for the time and context – in particular, excerpts from the slow movement of the C major Quintet, the A-flat Mass, the Eighth Symphony, the slow movement of the E flat Piano Trio, the A Major Piano Sonata, and the C-major Fantasie for Violin and Piano, and several songs (»*Die Krähe,*« »*Ihr Bild,*« »*Stürmische Morgen,*« »*Der Leiermann,*« »*Die Doppelgänger*«), but many sequences transpire in a protracted, wordless silence pregnant with the incidental aural textures and ambience of Schubert's world. »Some people protested that I did not use music all the time,« says Lehner. »So many movies about composers think you have to play music *all the time*. But no. You will note I only use it sparingly in the film, and only then when it is absolutely *meaningful*.« The myriad noises of the Viennese streets, the haunted whispers of the hidden grotto, the scratches of the quill pen and the buzzings of the flies at Schubert's desk convey meanings impossible to articulate in words. Indeed, with the exception of a handful of scenes, the viewer scarcely needs an English translation.

Mit meinen heißen Tränen is a milestone in the history of composer biopics and, more to the point, marks a radical shift in the nature of dramatizations of Schubert's life and milieu.[36] Yet, in 1989, three years after the first telecast of *Mit meinen heißen Tränen*, biographer Maynard Solomon could declare, »Thus far, biographers have not been able to provide even a provisionally convincing portrait of his personality, to delineate his obsessions, understand his familial and intimate relationships, and ultimately to glimpse some of the driving forces of his creativity« (206). Has Lehner's film, with its tarnished portrait of milieu and man and its revelations of sexual confusion and venereal disease, failed to enlighten us about Schubert's life and music? As an aural-visual dramatization, does it qualify at all as a responsible and factual biography, any more than its predecessors? Ultimately, has it failed altogether to overturn the »Blossom Time« stereotype? »There is no reason to suppose,« writes biographer Brian Newbould, »that, among those who cherish the world of fantasy-made-legend, the straightforward, *gemütlich* image of *Schwammerl* will not linger on into the twenty-first century« (407). We should ask ourselves if these Biedermeier stereotypes constitute any less valid a »history« than the revisionist versions. Christopher H. Gibbs reminds us that the »Blossom Time« operettas captured a side of Schubert – his music as unsurpassed melody and popular entertainment – that by the end of the nineteenth century had been hitherto overshadowed by his more substantial works: »Paradoxically, even as Berté trivialized Schubert's life and work,« writes Gibbs, »he resurrected the Biedermeier composer whose small-scale songs and dances first delighted Viennese audiences of the 1920s« (54). Critic Tim Page adds, »Perhaps it is both naïve and wrongheaded to blast the popularizers. Such is the prevailing mystery of the man that I wonder whether our ›marginalized‹ interpretation of his life is any more definitive than the sentimental portrait drawn by the Victorians« (n.p.). After all, notes commentator Alex Ross, certain aspects of the Biedermeier Schubert are true enough to confound the revisionists. A number of songs, including »Voluntary Oblivion« and »I Creep Around, Anxious and Silent,« rival even the effusions of the operettas. Moreover, continues Ross, »Fans of the dark side don't know what to do with documents incontrovertibly demonstrating that he liked to sit in coffeehouses and play skittles« (70). Doubtless the real fault lies in Schubert himself. Here was undeniably a man of profound psychological and sexual ambivalence who wrote surpassingly enigmatic music, fraught with the agony and joy of ceaselessly oscillating major and minor modes, frequently within the span of a single measure.

All of that notwithstanding, what Fritz Lehner's *Mit meinen heißen Tränen* does do is thrust Schubert into a relentlessly *physical world* of Vienna and its surrounding countryside, immersing the composer in a succession of sensuous moments, and leaving us intoxicated with a prevailing sense of strangeness and mystery. It places Schubert *in* and *of* his world, grounding – and then releasing – his sublime music from the tainted circumstances from which it sprang. In my opinion, in no other filmed dramatic biography of an artist has this miracle been equaled.

NOTES

1 The author wishes to thank Dr. Janet Wasserman, Founder and Executive Director of the Schubert Society of the USA, Richard Morris (editor of the SSUSA website, www.carolinaclassic.com/ssusa), and Professor Michael Saffle for their generous contributions to my research.

2 Author's telephone interview with Fritz Lehner, 7 January 2003, Vienna. Lehner is a native of Austria and studied filmmaking from 1970 to 1975 at the Film Academy in Vienna. He works for ORF television and freelances as a director. *Mit meinen heißen Tränen* was his first film about a composer. At this writing he is planning a film about Gustav Mahler.

3 »Gottfried Biedermaier« was the pseudonym of two writers, A. Kussmaul and L. Eichrodt. The poems originally appeared in the humorous weekly, *Fliegende Blätter*. The so-called »Biedermeier« period came to characterize the years in Austria between the Congress of Vienna in 1815 and the Revolution of 1848. Although a period of artistic fecundity, it was also a time marked by what Robert Waissenberger calls a »decline in taste.« He blames this in part on the Metternich system, »with total state control of all intellectual activity, which had the effect not so much of reducing the scope of artistic expression as of diluting its quality. People retreated within their own four walls, the better to escape notice. There, in the domestic circle, a favorite pastime was music, which, when it was not associated with ›revolutionary‹ lyrics, was quite beyond the philistine comprehension of the censorship authorities. [...] The result was a prodigious spread of amateurism during this period« (261-2). For a detailed discussion of the subject, see Waissenberger, *Vienna in the Biedermeier Era.*

4 For a detailed account of Kreissle's biography and its translation into English, see Gramit's »Constructing a Victorian Schubert: Music, Biography, and Cultural Values.«

5 The so-called »Unfinished Symphony« (the Eighth Symphony), popularly presumed to be the emblem of an unrequited love at the end of Schubert's life, was, as biographer Brian Newbould points out, a work that was composed six years before his death and

completed in sketch form. »Schubert's own attitude to the work was ambivalent, to say the least. He wrote it for no occasion, for no commissioner, for no individual or organization, but from the lifelong inner compulsion to write symphonies« (182). See also Newbould's discussion of the process by which he »realized« the symphony, in »On Completing Schubert.«

6 *Blossom Time*'s original Broadway cast included Bertram Peacock as Schubert, Olga Cook as Mitzi, and Howard Marsh as Schober. The show became a perennial favorite all over the United States, with at least one company appearing somewhere every year. Laufe reports that it proved to be a forerunner of the long-running Broadway musicals of the 1940s that adapted music by classical composers (Grieg in *Song of Norway*, Borodin in *Kismet*, and Bizet in *Carmen Jones* (13)). The song known as »Ave Maria« was in reality a *Lied* called »Ellen's Gesang« and was composed in 1825 to Canto III of Sir Walter Scott's *The Lady of the Lake*. Although it customarily figures in Schubert shows and films as a church hymn, musicologist John Reed observes that its emotion »is secular and aesthetic rather than religious« (217).

Wasserman has left us an account of a recent revival of *Lilac Time*, as performed at a summer workshop at the Mannes College of Music, on 30-31 July, 2003. See the SSUSA Newsletter, Vol. 1, No. 1 (2003): 3.

7 For more details about these films, see my *Composers in the Movies: Studies in Musical Biography.*

8 Wasserman has compiled an exhaustive list of films that either dramatize Schubert's life or at least feature his music in some way, either adapted for the soundtrack score or as an occasional performance in the diegesis. Contact her at the Schubert Society of the USA (e-mail: SchuSocUSA2003@aol.com). For a survey of the various incarnations of Franz Liszt in the movies, see my »The Truth in Masquerade: Images of Franz Liszt in the Movies.«

9 The *Schubertiaden* began in 1988 with the intention of publicly performing, in chronological order, all of Schubert's compositions. In successive years this agenda was complemented with a series of talks and lecture-demonstrations (including picture exhibitions and film screenings), such as »Perspectives on Schubert – from Weimar to Hollywood,« organized by Joseph Horowitz for the 1992 *Schubertiade*. At this writing, another project is underway to present all of Schubert's songs to the public. Pianist-musicologist Graham Johnson has gathered many young singers to record the entire literature on the Hy-

perion label. The project is now in its 40th release. See my »Graham Johnson: Immersed in Schubert.«

10 Robert Waissenberger alleges that many of the *Schubertiaden* »merely provided a cover for meetings that were not without a certain intellectual and political explosiveness« (256).

11 Author's interview, New York City, 12 September 1996.

12 With profound understatement, Maynard Solomon writes, »Not everyone will accept the evidence that I offer here« (103). The publication of Solomon's »Franz Schubert and the Peacocks of Benvenuto Cellini,« in which allegations of Schubert's homosexuality and its possible influence on his music, aroused a firestorm of controversy, culminating in a heated debate at the 1992 *Schubertiade* in New York City. »This kind of discussion is among the most important of our time,« wrote critic Edward Rothstein. »Musical analysis is becoming less abstract, and critical interpretation steadily draws the most innocent of compositions into the hothouse world of contemporary politics. [...] The only thing agreed upon, probably, was that Schubert's personality is not well understood« (C11). The allegations were countered later by two articles by Rita Steblin, »The Peacock's Tale: Schubert's Sexuality Reconsidered« and »Schubert's ›Nina‹ and the True Peacocks« and by Marie-Elisabeth Tellenbach in »One Man's Meat.« In »Whose Schubert?« Robert Winter examines this discussion and points out that Solomon's allegations merely extrapolate on evidence already available since Schubert's own time. As for the exact nature of Schubert's illness, Eric Sams in »Schubert's Illness Re-Examined« exhaustively sifts through the evidence pointing to typhus and syphilis. Sams suggests that syphilis was indeed the immediate cause of his death and that Schubert's own foreknowledge, as early as 1823, of his certain end »could hardly have failed to be afflicted and modified thereby« (21).

13 For useful introductions to the academy's recent endorsement of historical films, see George F. Custen, *Bio/Pics: How Hollywood Constructed Public History*; Robert A. Rosenstone, »The Dramatic Film as (Postmodern) History«; Pierre Sorlin, »How to Look at an ›Historical‹ Film«; Hayden White, »Historiography and Historiophoty«; and Simon Schama, »Clio at the Multiplex.« Primary texts examining the classical Hollywood studio system, include David Bordwell, Janet Staiger, and Kristen Thompson, *The Classical Hollywood Cinema: Film Style & Mode of Production to 1960* and Janet Staiger, ed., *The Studio System*.

14 For two recent accounts of the work of Ken Russell and Tony Palmer, see, respectively, my »Elgar's Ear: A Conversation with Ken Russell« and »Shostakovich's Fool to Stalin's Czar: Tony Palmer's *Testimony*.«

15 For a detailed overview, see my *Composers in the Movies: Studies in Musical Biography*.

16 The shortened version shown in America was known as *Notturno*. It consisted of two parts, »Love Has Lied« and »Winter Journey.« The critic for *Variety* applauded its »superb cinematography, topnotch production values and fine thesping [...]« (Pressbook). The script of the complete film, accompanied by commentary by Theophil Antonicek and Gerald Szyszkowitz and many photographs by Gabriela Brandenstein, is available in a German-language volume, *Franz Schubert: Mit meinen heißen Tränen*, Edition Wien, n. d.

17 Modern Schubert research really begins with three seminal works by Deutsch. His *Documentary Biography* appeared in Germany in 1914 and was translated into English by Eric Blom in 1946. Five years later Deutsch published his *Thematic Catalogue*. His *Memoirs* first appeared in a German edition in 1957.

18 »Der Film Fritz Lehners ist bestrebt, in diese Risse hineinzuleuchten, die Unterwelt des ›Schwammerl‹ aufzubrechen. [...] Es geht aber nicht um Dokumentation, sondern offensichtich darum, mit den Mitteln der Filmkunst in der Darstellung Schuberts etwas von dem im Beschauer wachzurufen, was ihn aus dem Schaffen des Komponisten in jenen Jahren angreifen sollte. Gewiß hat Lehner sich der belegbaren Fakten versichert und sie auch miteinbezogen. Aber für die eigentliche Botschaft des Films sind sie kaum mehr als Anhalts- oder Ausgangspunkte, von denen aus die einfühlende Phantasie des Künstlers ihre Anstöße nimmt.« Theophil Antonicek, »Was die Wissenschaft weiß,« *Franz Schubert. Mit meinen heißen Tränen: ein Film Fritz Lehner* (Vienna: Edition Wien, n.d.). Hereafter, *Tränen* for references to the published screenplay contained in the same volume. English translations in the text are not literal translations of the artificial Viennese in the screenplay, but rather give the essence of the contents.

19 »Verzeihung« (*Tränen* 24).

20 »Räch dich an der Hur. Und nicht an mir. [...] Sei froh, daß ich dich immer wieder mitgnommen hab, sonst wärst jetzt noch eine Jungfrau [...] Eins bitt ich dich noch: zieh nur ja recht schnell aus meinem Zimmer aus« (*Tränen* 36-7).

21 »Ich tät sehr gern mit euch reden wollen, Herr Vater.« [...] »Er war einmal Lehrer – mein Sohn – ein Franz wie ich – Compositeur – immer unterwegs – Wein – Dreck-

huren […] Neunzehn Kinder sind mir geboren worden. Zehn hat mir Gottvater wieder genommen. Und dich hat er am Leben gelassen« (*Tränen* 20-1).

22 »Vielleicht ist es besser, wennst es weißt. […] Mir wars ja auch nicht recht, aber soviel Geld kriegt man nicht immer. […] Wennst willst, leg ich mich die ganze Nacht zu dir. Lieb haben wir dich alle, hörst!« (*Tränen* 86)

23 »Ich bin Schubert. Ich bin Franz Schubert. Franz Schubert« (*Tränen* 87).

24 »Hast nichts Leichteres? […] So wirst es nie zu was bringen. […] Schau dir den Paganini an, den Strauss. Die sind gschickt. Den Lanner, Rossini. Der Paganini kriegt für ein Konzert mehr, als du in drei Jahr« (*Tränen* 105-6).

25 »Verbrennt meine Kleider. Er hat Angst, daß ich ihn oder seine Frau ansteck. […] Dich Spaun, mag ich, weil du kommen bist […] oder hast das nicht gwußt? Weichts mir nur recht schön aus. Viel mehr trifft mich, dass es euch vor meiner Musik graust« (*Tränen* 106-7).

26 »Paß lieber auf, daß dich nicht ansteckst […] Und der Ferdinand soll meine Wäsch verbrennen, wenns ihm Spaß macht […] Als obs Sterben das Schlimmste wär […] Traurig bin ich, weil ich nicht weiß, was mir wirklich fehlt. Ich glaub, daß es alle wissen, nur ich nicht. Ist es meine alte Pest? […] An Nervenfieber ist meine Frau Mutter gstorben« (*Tränen* 107-8).

27 »Eigentlich war ich nur neugierig darauf, was Sie sehen, wenn Sie zu mir rüberschaun […] Ich kann sehn, wenn Sie im Bett lesen […] Sie sind ein Dichter. Ich seh Sie auch schreiben […] Komponierens auch Musik zum Tanzen?« (*Tränen* 111-12).

28 »Tanzen mir […] Kommens. Komm. Willst mich ja« (*Tränen* 112).

29 »Aber tanzen gehn wir bald« […] »Sie brauchen mir ja nur zu winken. Ich seh Sie ja eh immer« (*Tränen* 113).

30 »Lieber Schober! Ich habe schon 11 Tage nichts gegessen. […] Sei also so gut, mir in dieser verzweiflungsvollen Lage durch Lektüre zu Hilfe zu kommen. Von Cooper habe ich gelesen: *Den letzten Mohikaner*, den *Spion* […]« (*Tränen* 116).

31 »Wie haben einen neuen Arzt, Franz, und der sagt, dass du wieder gesund wirst« (*Tränen* 127).

32 »Ferdinand, wennst noch mein Bruder bist, dann schick ihn weg, ich bitte dich. Hab ihn nie braucht, brauch ihn jetzt nicht […] Ferdinand, du warst mir immer der liebste […] Versprechen mußt mir auch, daß sie mich nicht so herrichten wie den Beethoven, dem sie zum Andenken die Haar abgschnitten haben, daß er dann ganz glatzerte Stellen

ghabt hat. Oder daß mir den Kopf abschneiden, wie es dem Haydn passiert ist« (*Tränen* 129).

33 Sogar die Pfeifen schmeckt dir schon wieder [...] [Schubert: »Ich spürs ja auch, wie es mir von Tag zu Tag besser geht« to which Schober replies:] »Hast grad so schön gsungen« (loc. cit.).

34 It is true that Schober kept rooms for Schubert when they were needed. Schober wrote texts to several Schubert songs, including »*An die Musik*,« »*Todesmusik*,« »and »*Am Bach im Frühling*;« and the opera, *Alfonso und Estrella*, on which they collaborated early in 1822. Schober is usually singled out as a regrettable influence on Schubert. As biographer Charles Osborne points out, »many of the Schubertians were distrustful of Schober's influence over their friend. In particular, they were distressed that Schubert had begun to ape the loose-living ways of Schober, who seemed to delight in initiating others into his own style of life, in which wine and women certainly played as large a part as song.« Newman Flowers acknowledges Schober's baleful influence over the man, citing a remark by Josef Kenner that Schober gained over Schubert »a lasting and injurious influence.« After Schubert's death, Schober went to Weimar, served as a secretary to Liszt for awhile, married later at age 60, and after the divorce died a recluse in Dresden. See Charles Osborne, *Schubert and His Vienna* (New York: Alfred A. Knopf, 1985), 90; and Newman Flower, *Franz Schubert: The Man and His Circle* (New York: Frederick A. Stokes Company, 1928), 136-7.

35 Brian Newbould writes that Josefa was Schubert's 13 year-old half-sister who, along with Ferdinand's wife Anna, nursed the dying composer in his last days (1997, 275).

36 Its influence on subsequent Schubert films is obvious. For example, scarcely a trace of the »Blossom Time« paradigm remains in two British biopics, Christopher Nupen's *Franz Peter Schubert: The Greatest Love and the Greatest Sorrow* (1994) and Peter Webber's *The Double Life of Franz Schubert* (1997), both of which openly acknowledge Schubert's sexual ambiguity and contraction of syphilis. Nupen's film adopts a semi-documentary approach to Schubert's last years, blending performances by Vladimir Ashkenazy, Andreas Schmidt, and the Bavarian Radio Symphony with recitations of Schubert's letters, allusions to paintings by Caspar David Friedrich, and dramatized sequences. Webber's film grimly and not very subtly dramatizes the circumstances surrounding Schubert's contraction of syphilis in 1822-1823. The film's formal structure intercuts scenes amidst the squalor of hospital scenes with Schubert (Simon Russell Beale) with events leading

up to the incarceration – his abandonment of teaching, his meeting with Schober (Jason Flemyng), episodes involving drugs and prostitutes, his brief infatuation with Karoline Esterhazy, and his collapse from the first syphilitic attack. Schober is portrayed as a leering hedonist who repeatedly mocks Schubert's timidity and frequently consorts with prostitutes. A recurring dream motif depicts Schubert lying recumbent in his hospital bed, his visage like a death mask, the soundtrack fraught with moaning sounds, accompanied by a stream of images of Schubert's night-shirted form advancing toward a coffin that holds his corpse.

REFERENCES

Antonicek, Theophil and Gerald Szyszkowitz. *Franz Schubert: Mit meinen heißen Tränen*. Vienna: Edition Wien, n. d.

Bordwell, David, Janet Staiger, and Kristen Thompson. *The Classical Hollywood Cinema: Film Style & Mode of Production to 1960*. New York: Columbia University Press, 1985.

Botstein, Leon. »Realism Transformed: Franz Schubert and Vienna.« *The Cambridge Companion to Schubert*. Ed. Christopher H. Gibbs. Cambridge: Cambridge University Press, 1997. 15-35.

Brendel, Alfred. »Schubert's Last Sonatas.« *The New York Review of Books* 2 February 1989: 32-6.

Custen, George F. *Bio/Pics: How Hollywood Constructed Public History*. New Brunswick: Rutgers University Press, 1992.

Davis, Peter G. »Out Week.« *New York* 17 February 1992: 59-60.

Flower, Newman. *Franz Schubert: The Man and His Circle*. New York: Frederick A. Stokes Company, 1928.

Gibbs, Christopher H. »›Poor Schubert‹: Images and Legends of the Composer.« *The Cambridge Companion to Schubert*. Ed. Christopher H. Gibbs. Cambridge: Cambridge University Press, 1997. 36-55.

Gramit, David. »Constructing a Victorian Schubert: Music, Biography, and Cultural Values.« *Nineteenth-Century Music* 17.1 (Summer 1993): 65-78.

Laufe, Abe. *Broadway's Greatest Musicals*. New York: Funk & Wagnall's, 1977.

Newbould, Brian. *Schubert: The Music and the Man*. Berkeley and Los Angeles: University of California Press, 1997.

_____. »On Completing Schubert.« *The Schubertian* 32 (July 2001): 6-13.

Osborne, Charles. *Schubert and His Vienna*. New York: Alfred A. Knopf, 1985.

Page, Tim. »Schubert, Revised.« *New York Newsday* 4 February 1992.

Reed, John. *The Schubert Song Companion*. New York: Universe Books, 1958.

Rosenstone, Robert A. »The Dramatic Film as (Postmodern) History.« *Revisioning History: Film and the Construction of a New Past*. Ed. Robert A. Rosenstone. Princeton: Princeton University Press, 1995. 202-9.

Rothstein, Edward. »Was Schubert Gay? If He Was, So What? Debate Turns Testy.« *The New York Times* 4 February 1992, C11.

Ross, Alex. »Great Soul.« *The New Yorker* 3 February 1997: 70-8.

Sams, Eric. »Schubert's Illness Re-Examined.« *Musical Times* 121.1643 (January 1980): 15-22.

Schauffler, Robert Haven. *Franz Schubert: The Ariel of Music*. New York: G.P. Putnam's Song, 1949.

Schama, Simon. »Clio at the Multiplex.« *The New Yorker* 19 January 1998: 38-43.

Solomon, Maynard. »Franz Schubert and the Peacocks of Benvenuto Cellini.« *Nineteenth-Century Music* 17.1 (Summer 1993): 193-206.

Sorlin, Pierre. »How to Look at an ›Historical‹ Film.« *The Historical Film: History and Memory in Media*. Ed. Marcia Landy. New Brunswick: Rutgers University Press, 2001. 25-49.

Staiger, Janet, ed. *The Studio System*. New Brunswick: Rutgers University Press, 1994.

Steblin, Rita. »The Peacock's Tale: Schubert's Sexuality Reconsidered.« *Nineteenth-Century Music* 27.1 (Summer 1993): 5-33.

_____. »Schubert's ›Nina‹ and the True Peacocks.« *Musical Times* 138.1849 (March 1997): 13-19.

Tellenbach, Marie-Elisabeth. »One Man's Meat.« *Musical Times* 141.1870 (Spring 2000): 50-2.

Tibbetts, John C. *Composers in the Movies: Studies in Musical Biography*. New Haven and London: Yale University Press, 2005.

_____. »Elgar's Ear: A Conversation with Ken Russell.« *Quarterly Review of Film and Video* 22.1 (January-March 2005): 37-49.

_____. »Graham Johnson: Immersed in Schubert.« *The American Record Guide* (November-December 1999): 22-5.

_____. »Shostakovich's Fool to Stalin's Czar: Tony Palmer's *Testimony*.« *Historical Journal of Film, Radio, and Television* 22.2 (2005): 173-96.

_____. »The Truth in Masquerade: Images of Franz Liszt in the Movies.« *Liszt the Progressive.* Eds. Hans Kagebeck and Johan Lagerfelt. Lewiston NY and Queenston Ontario: The Edwin Mellen Press, 2001. 209-22.

Waissenberger, Robert. *Vienna in the Biedermeier Era.* New York: Mallard Press, 1986.

White, Hayden. »Historiography and Historiophoty.« *The American Historical Review* 93.5 (December 1988): 1193-9.

The Creator, the Plunderer, the Object and its Lovers: Austrian Global Musical Crossovers as a Floodgate of Significations across Times and Spaces

CORNELIA SZABÓ-KNOTIK

INTRODUCTION

Allusive music, that is, music which refers to other pieces of music and puts already-existing pieces into a different context by quoting them in a recognizable way, is typical of the enlightened way in which this art form was handled. Through the allusive approach, which began at the end of the eighteenth century, music and its creators became objects of esteem and veneration. This process was furthered for a number of reasons in the following decades when the discourse of composers became a more frequent phenomenon and romantic ideas together with historicism meant an increased attention to folklore, to chant as the basis of »true church music« and to music of bygone periods in general (research began digging backwards and reached Gothic times towards the turn of the century). As a result, the nineteenth century became a period in which musical quotations were frequently used and music about music flourished, labeled with different names, such as variations, transcriptions, paraphrases, and phantasies.

When towards the end of the nineteenth century the opposite concept of the original work of genius was made absolute, an appreciation and understanding of this canon vanished and these pieces became regarded as a creative weakness of the composer in question (prominent cases: Liszt, Mahler). In our visually dominated culture, the view that an original piece by a genius is the only form of musical art is no longer common, but allusive music (referring to music, to literature or to the arts) is attracting increasing attention, which I regard as a symptom of the fact that we recognize ourselves in these pieces, meaning that we find parallels to current developments in and attitudes towards music. These developments and attitudes are strongly influenced by or even caused by the outcome of the twentieth-century technological revolution: sound recording and broadcasting led to the contemporary situation, in which music from all over the world, from all kinds of genres and many centuries – in short, a vast »ocean« of music – is virtually accessible at any moment anywhere in the world. This is a historically unique situation, which poses special challenges to contemporary production. On the one hand, procedures taken from the arts, such as montage and collage, are applied in the search for musical progress; on the other, the differences between the popular and the artificial seem to blur as both kinds of music work with similar means, in production (composition) as well as in marketing, and both help themselves to the above-mentioned ocean of pre-existing music which (cross-over being the cue) they quote, arrange, combine and assemble in remixes, cover versions, and collages.

It is the aim of this paper to discuss two recent examples from »Austria, the Land of Music« (as this country's identity-construction will have it), which illustrate some interesting twists in the interplay of the local and the global in and through music, and to consider how the network of intra-musical references they depict places them as part of a process of cultural flows constantly producing change and exchange of images and significations.

TWO EXAMPLES

1) The Mozartband

The Mozartband consists of ten musicians and singers of different training and experience in classical as well as in popular music and is directed by Wolfgang Staribacher, who has been part of Hubert v. Goisern's Neue Volxmusik ensemble.[1] They also form a smaller group called Mozart Bandits, enhancing their options as it can perform on smaller stages (also, for less money). The Mozartband's first performance took place at the 1999 *Wiener Festwochen* with the premiere of *König des Glücks* (*King of Good Fortune*), an arrangement of W. A. Mozart's *Il Re Pastore*.[2] They then went on tour and produced a CD called *Soul*, which contains some tracks from the *Festwochen* production.[3] In 2003 one of their Viennese concerts was recorded on video,[4] which adds showy staging that clearly establishes the ensemble's position in popular culture and also transfers the corresponding image of Mozart typical of the ensemble's web-design[5] as well as the CD's cover-title, on which a long-haired male face, whose arms rest on an accordion, sticks out its tongue, in imitation of pop stars and Albert Einstein's iconic late photograph.

The plot of *Il Re Pastore* intertwines two stories. The story of Alexander the Great (who comes to free the town of Sidon from an usurper whom he wants to replace with a humble young shepherd who turns out to be the real heir to the throne) is told by Alexander's Phoenician adviser together with this shepherd/king's love story (he loves a noble girl but should marry the usurper's daughter). The opera's second act contains an aria (*Se vincendo*) in which Alexander, after having decided to leave Sidon after the shepherd has ascended to the throne, expresses his belief that he has fulfilled all of the wishes, leaving just peace and happiness behind him. The aria's original version brings the text's two pairs of verse twice (see Appendix) and is characterized by much virtuoso singing, closing accordingly with a concerto-like cadenza. The Mozartband's version (CD track 8) is somewhat shorter, often keeping the accompagnato figuration from the original and – leaving out the cadenza – reserving coloraturas for the repetition of the whole text. It prominently stages the special, chamber music-like sound of mixed instrumentation with a female, opera-trained voice, and this »clash of sounds« creates a special effect throughout the recording.

Another example which is especially interesting concerns an original related to a tradition of significations mediated through various stagings and is the only »all-time Mozart favorite« the ensemble has thus far recorded. The title-hero's so-called *Cavatina* from *Le Nozze di Figaro* has in the last decades been interpreted ambiguously on a scale between a jealous servant's witty parody of a serenade to an almost undisguised threat to feudalism anticipating the French revolution's dawn. Mozart's setting first evokes the sounds of a serenade with a plucked accompaniment to the voice during the first verse, then (*Se vuol venire nella mia scuola*) the orchestra enters and increases in dynamics and intensity at the word *saprò*, after which it quickly calms down to the words *piano, piano* – the threat of rebellion only hinted at. The verses from *meglio arcano* to the end form a middle part, after which the beginning is repeated. That's the original. The Mozartband does more and does it more clearly, their version being the only recorded piece on the CD that is made longer. Starting with an introduction of accompanying figures on the guitar (in the original singing starts right away), the first verse alludes in their interpretation to what could be seen to be the contemporary equivalent of a serenade, namely, a romance. This romance consists of the ongoing guitar accompaniment and the male voice of a blues singer instead of opera-style intonation, with instrumental tremolos imitating mandolina sounds at the word »chitarrino.« Percussion adds regular beats when the first verse is repeated and this time *sapró* [...] *ma piano* brings full rock drive with electric guitars as a contemporary version of rebellious music, but with a plugged bassoon as the most prominent and unusual voice, playing an instrumental interlude to the next phrase (*se vuol venire*). This verse is also repeated and forms the culmination of the song, to which the singer's voice contributes together with hammered basses and e-guitar figures. The original's intermediate part is left out, and the song ends with a coda of idyllic »Mozart cliché« phrases, again seemingly relating to the original but being an addition.

2) The Global Kryners

The Global Kryners is a six-member band, which encompasses such different fields of musical practice as commercial folklore, jazz, pop and brass band. As in the case of the Mozartband, the name stands for the band's musical commitment as expressed on its website, which is to revive the sound of a variation of the commercial alpine folklore that was successful in the 1950s: the so-called oberkrainer style (Oberkrain being the old name for the region of upper Slovenia where the first ensemble came from), to »translate it into a serious genre of world music« and »to prove that every song, every melody of jazz, pop and classical music can be played in krainer style.«[6] The presentation on the web plays with the clichés of commercial folklore in the sense of Oktoberfest-nostalgia,[7] as does the CD's cover, which shows the outlines of five naked musicians with traditional instruments against a clichéd landscape of cutely clouded blue sky and a yellow spotted meadow in a pose traditional for folklore or popular associations (be they music bands or sport teams). A female figure in a dirndl is in front, the folklore costume in its commercialized variant mainly framing her prominent cleavage.[8]

It is up to the consumer to understand this presentation as a parody of clichés, and the same is also true of the sound. In its beginnings the oberkrainer sound[9] meant a new interpretation of traditional dances like the polka, waltz and march by means of instrumentation. Its jazz guitar was the most prominent difference to folklore usage. Beat rhythms lent the music a modern sound, and the melodies played on clarinet and trumpet were shared and accompanied by an accordion in a new way that was taken as being more dynamic.

The songs presented on the CD are mostly Anglo-American pop-hits from the sixties (e.g. *Something Stupid*) to the present (e.g. *Sex Bomb*) interspersed with some golden oldies from the early decades of the twentieth century (*Stardust, Somewhere over the Rainbow*). The arrangements rely to a great extent on the female singer's virtuoso voice with its broad range of expressive means. The fact that an oberkrainer band's sound is so different from the usual instrumentation also creates a considerable effect and makes the music sometimes light and funny, sometimes more emotionally intense. The fixed patterns of the original oberkrainer music with which the well-known pop-songs are combined are especially apt for medley practices of performance and have therefore mainly the function of introduction figures and transition passages. Used for instrumental intermezzos between the verses of the songs the given pop-tune's well-known melodies are worked (melodically, rhythmically) in a spirit of inprovisation that stages the instrumentalist's virtuosity.

Lady Marmalade was written by Bob Crewe and Kenny Nolan in 1974 and became this team's biggest hit. Recorded first by a studio group called The Eleventh Hour and then by Patti Labelle, it was covered in 1998 by the female British pop-group *All Saints* and then again in 2001, when it was produced by Missy Elliot with Christina Aguilera, Lil'Kim, Mya and Pink for use in the movie *Moulin Rouge!* (2001, dir. Baz Luhrmann) and won an MTV Video Music Award.[10] The Global Kryner version (CD track 2) starts with a trombone solo and an oberkrainer pattern, a kind of polka, presenting a whole verse instrumentally and introducing the cool vocal first lines (*hey sister, go sister*). While the voice keeps the original disco sound, the instruments (mainly trombone and accordion) start to weave their first oberkrainer-like version into the accompaniment (in the trombone's accompanying figures). The main feature of the musical mix where disco and oberkrainer meet most closely is rhythm. The first quotation of the famous chorus of *Voulez-vous coucher avec moi ce soir* adds the instrumentalists' male voices and leads into another folkloristic instrumental verse that again forms a transition to the vocal verse. This time the female voice joins the chorus and both continue singing with the instruments for a final repetition of the folkloristic section of which the last part is again merely instrumental, a typical oberkrainer cadenza. In terms of cultural exchange, this represents a tricky mixture of multiple interaction as the oberkrainer style is (was) already a cross-over between folklore practices and jazzy harmonies, an adaptation of local music to international pop culture's influences.

More traditionally written pieces like Billy Joel's 1978 *Honesty* (track 3) adapt so smoothly to the arrangement that the listener can more or less maintain an impression of unbroken musical wholeness instead of a clash of styles, while the Global Kryner's version of the equally sentimental 1968 John Lennon–Paul McCartney golden oldie *Hey Jude* (track 12) relies more on combinational effects, using the allusion to the verse for surprising musical twists and additions. Starting the tune quite simply with vocal and guitar, at the third verse (*And any time you feel the pain*) clarinet and trombone enter softly, the clarinet as second voice to the singer and the trombone profoundly marking the bass. For the last line of this verse (*By making this world a little colder*) the melancholically sounding accordion is heard with the melody's characteristic cadenza, used to bridge two verses, and the guitar very softly imitates this phrase, which is heard as the song's fading end, fitting to its general emotional content, an impression that is suddenly swept away by a prominent trumpet signal. This trumpet signals the entry of the whole band playing the song's polka-like oberkrainer version of its second part. After the first verse of this part (*So let it out and let it in*), the main voice is taken over by the accordion now showing its virtuosity, with guitar, clarinet and trombone accompaniment. When for this part's second verse the voice joins in with its close-to-the-original pop-style version, the combination of the two musical worlds presented so far in this recording provides an occasion for playful interaction.

ON THE INTERPLAY OF THE LOCAL AND THE GLOBAL IN THE PROCESS OF ACTUALIZATION

In discussing these examples I have to note that my criteria for evaluation are also a mix of traditions: partly popular culture, which is the field the Mozartband presents and markets itself in but, on the other hand, the application of criteria like virtuosity, creativity and originality of interpretation to characterize these recordings corresponds to the discourse of art music (cf. the Mozartband's bassoon solos, some violin duet passages, the Global Kryner's qualities of instrumental solos, of vocals, etc.). In this respect the recording has a distinct touch of »elite« as well as of commercialized entertainment.

It seems that the keyword for such an approach to music is *actualization*, which, remarkably, is the way in which at the beginning of musicological interest (before the discipline itself was institutionalized) pieces beyond the canon of contemporary production were received.[11] Through its origin the approach of actualization can be recognized as another hint about the influence social changes in a globalized world have had on Western music culture that is for many reasons gradually becoming an »endangered tradition« in certain ways.

The Mozartband's procedures seem to aim at reproducing the classical model's meaning, but garnish it with recent elements of the composer's image as a symbol of youth culture (*Amadeus*[12]). This combination is also part of the band's web presentation, a presentation that is interesting for the care it takes in inventing its own tradition and for its over-all classy, art-like character of a collection of »serious« information.[13] The description of the place where the band, together with manager and dramaturgical advisor Claudia Weinzierl, works has undertones of poetic nostalgia for all its special flair and for the creative community life it is said to represent: the »old villa at the outskirts of Vienna« alludes at the same time to images of a certain splendor (Vienna, the capital of the Habsburg Empire) and of idyllic nature (the Vienna woods); reflecting »chaos« and »power filled with anarchy,« it seems to be a magical place into which »the idea, the project, the soul of the Mozartband« has wormed itself, a place where »music, talks and laughter« add to the work done. Such lines invoke a kind of a Nashville community spirit, a blissful unity of creativity and life which – as not even a reference to the »lonely pot of simmering goulash« in the kitchen is forgotten – seems to be the opposite of isolation and alienation from work, the positive utopia of a life in wholeness, containing the sublime as well as everyday needs.[14] This allusion to the post-1968 phenomenon of alternative youth culture is strengthened by the elements of popular music on stage and in the arrangements. From the big, impressive amplifiers and loudspeakers we see on the video to the instrumentation, the technical equipment and musical design quote early 1970s classic rock, as do the hairstyles. The singers' costumes call to mind the fancy look that characterized contemporary youth culture's dress codes like the hippie or the Amadeus look.[15] Keeping with the artsy image, the timetable of acts on the website shows that the band mainly performs at special, classy occasions such as art-festivals and does not play any pop-music adaptations of Mozart pieces but rather takes on work in the concentrated form of projects, such as *Gaukler Sonata,* their 2003 cooperation with circus artists, or their new 2005 CD, which obviously means giving up performing for months of studio-work.

The Mozartband can thus be described as a very special combination of elements representing at the same time actualized tradition (classical music) and historicized contemporaneity (popular culture). Its music is yet another cross-over of so-called international popular music with the regional, i.e. with local color music, in other words another variant of *Neue Volxmusik*, one which dares to take classical music, music by W. A. Mozart, as a signifier of the local.

The description of this music on the band's website is entitled »Master Amadeus, the Folklore Musician« and begins as follows:

> A seemingly harmless standard band: percussion, E-bass, guitar, accordion, violin. But looking closer one is completely startled. The rhythm is grooving, the bass drumming and the eyes of the fans are gleaming. But the sound has a special kick, sounding from afar like Mozart. Five minutes later it becomes clear: this IS Mozart. But completely different. As though the master were still alive and writing his songs as *neue Volksmusik*: saucy, pungent, electrifying, completely up-to-date and rousing for every audience [...][16]

Leader Wolfgang Staribacher claims that at the beginning of this project stood his intense striving for a suitable way of performing Mozart's music on his instrument, the accordion, a striving initiated by a moment of fascinated listening of his, which occurred when he first heard Mozart's 12th Symphony (KV 110, written in 1771 for Salzburg), recorded by the Prague Chamber Orchestra with Charles MacKerras conducting. This recollection recalls images of unspoilt, idyllic life similar to those intended by or alluded to both by youth culture's social utopias as well as the center, the heart of what Mozart's music stands for: »I will never forget this moment: the peak of summer in Italy, a tent, walkman, surrounded by two dozen Mozart CD's. Unbelievable how he partly was grooving close to rock'n roll.«[17]

For two reasons Mozart's music is not a deliberate choice for this crossover but an almost logical one. First, there is the technical argument. Like all classical music in the precise sense of the word, i.e. a certain type of end-of-eigthteenth-century art music, it shares patterns of composition technique with folklore, namely the diatonic tone system, motifs and accompanying figurations built from triads and a regular two-plus-two ordered bar structure of forms. An especially good example of the similarity of folklore and Mozart's music in terms of composition material is the Mozartband's version of an excerpt from his 10th Symphony (KV 74, written 1770 in Milan and planned as the introduction to another one of his early operas, *Mitridate Re di Ponto*, CD track 1).

In addition to these material similarities, there is the matter of cultural paradigm, which gives meaning to the Austrian local being represented by Mozart's music instead of folklore. Mozart's music is a special musical signifier of Austrian identity, and this to such a degree that the composer can be called a patron saint of Austria's Second Republic[18] – a construction whose origins provide another remarkable parallel to the Mozartband's presentation and signification. It was at those times, in the months and years after World War II, when shortages and restrictions were the preconditions for an opera ensemble soon becoming one of the most prominent ambassadors of Austrian (music) culture at home as well as abroad, an ensemble living and working in close contact and continuity, an ensemble that, under the name »Mozart Ensemble« was, a mere decade later, glorified as the personification of a golden age of opera history in Austria.[19]

The Global Kryner's procedures, on the other hand, represent an interplay of the global and the local in the field of openly commercialized music. The listener's attention is attractively kept busy with expectations broken and fulfilled not by the process of mixing »art« with »the popular« but by the process of combining two kinds of popular entertaining music having the same historical origin (the second half of the twentieth century) but shaped by distinctly different local and social traditions. The oberkrainer style breaks the foreseeable regularity of the popular song together with the original's tone, while it is this tone that disturbs the impression of folklore kitsch. At first glance it seems that the Global Kryner act is yet another model for the reproduction of meaning in a traditional product similar to Bertold Brecht's so-called alienation effect. It is the global, the all-too-well-known popular tune which is preserved from trivialization by the originally commercialized, trivial local that is – I believe due to its outdatedness – used as an unexpected ingredient for an adaptation that regains the listener's attention and thus seems to recoup the popular tune's trivialization, transforming the all-too-well-known into something unusual. But this is just a first impression. Listening more often and more carefully, the arrangement's effect is produced rather by matters of which the commercial effect is well-known, such as instrumental and vocal virtuosity and easy-listening tunes.

The website is mainly commercially oriented, with no attempt to invent any myths and no pathos, but it has, like the CD cover mentioned above, some allusions to popular, non-urban humor. The return-button is shaped, for example, like a deer's head (German: *Hirsch*) – a traditional hunter's trophy[20] – and the corresponding inscription *Hirschickmaeichzruck* (here we send you back) alludes to a popular pun in which a phrase starting with the word »*Hirsch*« transforms into suggestive words when repeated quickly. The dense schedule of performances seems to prove by the selection of acts that the band's obvious success is almost exclusively taking place on the same stages the oberkrainer have helped to make popular, i.e. the world of so-called commercial folklore for mass entertainment in beer tents. Thus the Global Kryner have appeared at the big New Year's program of German-speaking television's best-known folklore show, the so-called »*Sylvesterstadl*« 2004/05 and they submitted a contribution to, and was Austria's representative at, the 2005 Eurovision Song Contest, during the broadcasting of which the winner is selected by phone-polls in all participating countries.

The oberkrainer adaptations of popular tunes are, therefore, not a critique of trivialization but an affirmation. The application of a performing style to a different style of music adapts it to the environment the performing style is coming from. And this is basically a logical consequence of a generational shift in its audience. The popular songs adapted are important for the generation now in its 50s and 60s,[21] which means that their transformation to the world of commercial folklore is intended to increase the popularity of commercial folklore, which has always consisted of older generation's music. From this perspective the Global Kryner simply matched a marketing strategy for popular brass music to a new audience. They did not invent the musical idea of adapting oberkrainer style to a different repertory, which was already done back in the 1970s by a band called Alpen-Oberkrainer, and they are not the only band doing such a thing but are currently part of a broader movement in commercial folklore.[22]

1) W. A. Mozart / P. Metastasio: *Il Re Pastore*

2nd act, N. 9 Aria:

Se Vincendo vi rendo felici / Si partendo non lascio nemici
Che bel giorno fia questo per me!
De'sudori, ch'io spargo pugnando / Non dimando più bella mercè

(If by winning I make you happy / If in departing I leave no enemies
what a wonderful day this will be for me!
For the sweat that I shed in battle / I ask no more reward.)

2) W. A. Mozart / L. da Ponte: Le Nozze di Figaro

1st act, N. 3 Cavatina:

Se vuol ballare, Signor Contino,
il chitarrino le suonerò.
Se vuol venire nella mia scuola,
la capriola le insegnerò.
Saprò […] ma piano,
meglio arcano
Dissimulando scoprir potrò.
L'arte schermando, l'arte adoprando,
di qua pungendo, di là scherzando,
tutte le macchine rovescierò.

(If my dear Count, you feel like dancing,
it's I who'll call the tune.
If you'll come to my school,
I'll teach you how to caper.
I'll know how […] but wait,
I can uncover his secret design
more easily by dissembling.
Acting stealthily, acting openly,

here stinging, there mocking,
all your plots I'll overthrow.)

NOTES

1 Cf. Barbara Boisits, »Austria's *Neue Volxmusik:* The Sound of the Global Village?« in *Reverbertations: Representations of Modernity, Tradition and Cultural Value in-between Central Europe and North America,* eds. Susan Ingram, Markus Reisenleitner, Cornelia Szabó-Knotik (Frankfurt am Main: Peter Lang, 2002), 105-12.

2 This is a so-called *festa teatrale* Mozart had written to one of the very popular texts by Pietro Metastasio for the visit of Archduke Maximilian, the youngest son of Empress Maria Theresa and later Bishop of Cologne, to Salzburg.

3 *Mozartband Soul,* lawine 2001, 07243 811138 2 0, www.blancomusic.de.

4 The video was directed by DORO, the team of Rudi Dolezal and Hans Rossacher, who rose to prominence on account of their work with international pop stars and also made Falco's successful *Rock me Amadeus* video.

5 www.*Mozartband*.at, accessed summer 2004.

6 »Sechs MusikerInnen aus den unterschiedlichsten Ecken sind nun angetreten, a) den Oberkrainersound aus der Bierzeltseligkeit in ein seriöses World-Music-Genre zu übersetzen, b) den Spielwitz dabei nicht zu verlieren und c) zu beweisen, dass jedes Lied, jede Melodie aus Pop, Jazz und Klassik verkrainert werden kann« (www.globalkryner.com, accessed summer 2004).

7 Cf. James Deaville, »*I' bin ein Kitchenerer:* Music and Identity in North American Oktoberfest Practices« in *Ports of Call: Central European and North American Culture/s in Motion*, eds. Susan Ingram, Markus Reisenleitner, Cornelia Szabó-Knotik (Frankfurt am Main: Peter Lang, 2003), 155-71.

8 This iconographic setting seems to represent the band's logo, its corporate identity as it is repeated not only on the website but also on posters indicating live-acts.

9 Cf. article »Oberkrainer-Stil,« in *Österreichisches Musiklexikon*, vol. 4, ed. Rudolf Flotzinger (Vienna: Verlag der Österreichischen Akademie der Wissenschaften, 2005), or: www.musiclexikon.ac.at (database usage to be paid).

10 Cf. www.songfacts.com, accessed autumn 2004.

11 A prominent example of such an actualization is also related to Vienna and to Mozart, who for a so-called »historical performance« of this kind had to change some of the instrumentation in Handel's Messiah, also adding two clarinets (instruments popular at Mozat's time).

12 For information on the role and effects of Milos Forman's award-winning movie, see my book *AMADEUS. Milos Formans Film als musikhistorisches Phänomen* Grazer Musikwissenschaftliche Arbeiten, vol. 11, ed. Rudolf Flotzinger (Graz: Akademische Druck- und Verlagsanstalt, 1999).

13 There are descriptions of all the elements taken into account for the fusion with Mozart's music, paragraphs entitled as follows: »The analysis of Mozart's musical language«; »improvisation and notation of the original«; »improvisation, amplification and the Mozartband's sound design«; »the Mozartband's creative way of working.«

14 »Claudia Weinzierl ist es, die die Mozartband mit all den schrägen Zaunreitern, deren besondere Talente nicht in Schubladen hineinpassen wollen, zu einem Ganzen gefügt hat. Die Dramaturgin und künstlerische Managerin der Mozartband zeichnet nicht nur für die Konzeption ihrer Inszenierungen verantwortlich, sondern ist Zentrum und menschlicher Kern all dieser Verrückten. Untrennbar mit ihr und Wolfgang Staribacher verbunden ist ihre Altbau-Villa am Stadtrand von Wien, in deren Inneren das ›Labor,‹ die Seele, das Herz der Mozartband brodelt, schlägt und pocht. Ein Ort der kreativen Energie. Hier haust das Chaos, so scheint es, eine Kraft voll purer Anarchie. Räume, erfüllt von Musik, Gelächter und Gespräch und jeder Menge ›Schmäh‹. […] Auf dem Herd in der Küche, einsam vor sich hin köchelnd, das Gulasch im Topf. Die Idee, das Projekt, die Seele der Mozartband, all das hat sich in dieses Gebäude eingenistet, hineingefressen, verankert. Für die Aufnahme der Schallplatte SOUL, wurden die Wände der Villa teilweise durchbrochen und miteinander verkabelt. Selbst das Klo wurde nicht verschont. Der Fremde, der diesen Kosmos betritt, erhofft sich die Offenbarung und findet – Menschli-

ches, allzu Menschliches. Nichts zeigt sich so wie man glaubt und doch ist alles erfüllt von dem, was man sich erträumt.«

15 Forman's title hero with his colored wigs was used as a model for dresses imitating a kind of neo-baroque style with frills and velvet bows to which extravagantly shaped shoes were to be added. This so-called »Amadeus look« was a great success from the boutiques on London's King's Road to the audiences at the Salzburg Festival, to department-store collections and dressmaking-patterns in home and garden magazines.

16 »Eine scheinbar harmlose Standard-Band: Schlagzeug, E-Bass, Gitarre, Akkordeon, Geige. Aber wer näher kommt, fällt aus allen Wolken. Zwar groovt und fetzt der Rhythmus, drummt der Bass und glühen die Augen der Fans. Aber der Sound hat einen ganz besonderen Kick, klingt von fern nach Mozart. Nach fünf Minuten ist dann klar: Das ist Mozart. Aber ganz anders. So als würde der Meister noch immer leben und seine Songs für die neue Volksmusik schreiben: frech, beißend, mitreißend, völlig *up to date* und für jedes Publikum zündend. Die Mozartband heißt dieser 1995 gegründete und seither immer erfolgreichere Radikalverein. Treibende Gestalt ist Akkordeonist Wolfgang Staribacher.«

17 »Die Idee kam mir, als ich zum erstenmal Mozarts 12. Symphonie hörte, 4. Satz, Prague Chamber Orchestra mit Charles MacKerras. Ich werde diesen Moment nie vergessen: Hochsommer in Italien, Strohzelt, Walkman, rund um mich zwei Dutzend Mozart-CDs. Wie der da in den frühen Symphonien teilweise am Rock'n'Roll vorbeigroovte, fand ich unglaublich. Ich begann, ein paar Allegretto-Sätze auf meiner Ziehharmonika zu spielen. Das ging überraschend gut. Mit den Helikonbässen meines Akkordeons konnte ich, nachdem ich eine neue Grifftechnik entwickelt hatte, die steady Achtelbässe gut swingen. Das Musetteregister im Diskant deckt die Geigenlage perfekt ab und liefert einen breiten, orchesterähnlichen Klang. Ich hörte nur mehr Mozart und begann langsam, seine Musik zu verstehen. Jetzt konnte ich mich auf die Suche nach Mitmusikern machen. Das hochgesteckte Ziel: Mozart zu spielen mit Respekt vor seiner Tonsprache und seinen Melodien, aber mit allen Freiheiten von Form, Arrangement, Groove und Sound, die das Original zuläßt, vielleicht sogar braucht.«

18 Cf. Gert Kerschbaumer, »Mozart – ›Arisch‹. Die politische Verwertung im Dritten Reich und die ›Austrifizierung‹ nach der Befreiung,« in *Das Phänomen Mozart im 20. Jahrhundert. Wirkung, Verarbeitung und Vermarktung in Literatur, Bildender Kunst und in den Medien* (Wort und Musik, Salzburger Akademische Beiträge N. 10), eds. Peter Csobádi, Gernot Gruber, et al (Salzburg: Verlag Ursula Müller-Speiser, 1991); also cf. my paper: »Mozarts Zopf und Schuberts Madonna: Barockisierungen im Aneignungsprozeß musikalischer Leitfiguren in Österreich,« Paper given at the International Workshop »Barock – ein Ort des Gedächtnisses. Ein Interpretament der Moderne/Postmoderne« in Rome, September 2004 (to be published in the proceedings of that conference).

19 Cf. my article: »Selbstinszenierung und Handelsbilanz. Die (Re)Konstruktion Österreichs nach 1945 mittels Musik,« in *Musik-Wissenschaft an ihren Grenzen. Manfred Angerer zum 50. Geburtstag,* eds. Dominik Schweiger, Michael Staudinger, and Nikolaus Urbanek (Frankfurt am Main: Peter Lang, 2004), 355-82.

20 Related to the world of hunting, mounted sports deer heads are a fixture of this lifestyle. The Global Kryner made more prominent use of it in 2005 as the logo of their website and on the cover of the single recording of their Song Contest submission, where it forms the lower part of a girl's bikini.

21 Cf. my article »Mythos Musik in Österreich: die Zweite Republik,« in *Memoria Austriae I. Menschen, Mythen, Zeiten,* eds. Emil Brix, Ernst Bruckmüller and Hannes Stekl (Vienna: Verlag für Geschichte und Politik, Oldenbourg Verlag, 2004), 243-70.

22 I thank my collegue Christian Glanz from the University of Music and Performing Arts Vienna for this information.

My Beloved Landscape:

Naming the Self in the Balkans

SRDJA PAVLOVIC

The eternal silence of these infinite spaces fills me with dread.
Blaise Pascal

The title of this contribution implies conflict and a subtle but nonetheless everlasting struggle between spaces and identities, between our immediate surroundings and us. It spells out the desire to own, and it also highlights the contested nature of our claims to the highest spot in a hierarchy of belonging, which could be summarized by the proclamation: We were here first! It is all about an orchard and an apple tree, and the mandatory presence of a serpent. It is all about a field in the Old Testament where Cain rules.[1]

It could also be argued that such conflict is a figment of the imagination and that things are the other way round: that the relationship between identities and spaces, which I am decoding as contentious and confrontational, is in essence harmonious and interconnected. Because of this interconnectedness, space demands that, with each step we take and each time we move, we conceptualize ourselves in relation to it. We cannot exist without locating ourselves, and simultaneously subjectifying and positioning ourselves. In spite of the fact that we have first to learn the rules of space/territory and adjust to them before we can act on our impulse to change and/or appropriate it, some might even maintain that it is people who define space by naming it; thus, we are the ones who really have the upper hand. Vanity and the will to possess seem to be main features of human history.[2]

Indeed, a name is a description with historical connotations, personal meanings, and cultural resonances drawn from our conscious environment but reflective of our subconscious selves. Every space/place has its own stories of spirits, demons, tribal fires, and ancestors, which through enfiguration we mould into history. While naming space we are appropriating such stories and locating ourselves within history, thus, claiming ownership to all three: space, story, and history. In naming, whether through everyday usage or idiosyncratic messaging, we express both the power of the past and the importance of our present activity. In naming, we are free to choose (are we really?) to be the same as others in our world, or to be different. The question is how we are to use such a choice, and through what kind of language it should be expressed. Should our choice be an expression of our memories about the past and our sense of belonging to a collective, or should our approach to space/place/territory/landscape and our relationship with it always remain an experience of the first kiss: new, unique, and detached from the past? Should we resort to a native tongue, or should we use Latin to tell our stories and name the space *insula nova undecesimo die Junius nubibus revelata*?

Following Jacques Lacan's observation that »no language can speak the truth about truth,« it would be naïve for us to believe that naming space and appropriating it by imposing on it memories of our ancestors could endow us with anything but a temporary and unstable sense of ownership. If anything, history teaches us that naming space does not always guarantee the eternal existence of peoples or even memories of them, let alone provide a safe harbor for our fragile identities. The different names given in different times to spaces/cities/towns in central Europe and the Balkans remind us of how brief were efforts in appropriating space through naming it according to tribal, ethnic, national, or religious criteria. Kolozsvar used to be Klausenberg, but now it is called Cluj Napoca. The town once known as Hermannstadt is now softly referred to by Romanians as Sibiu. What the Szeklers call Szekelyudvahely is to Romanians Odorheiu Secuiesc, Roman Apulum, their Alba Iulia a.k.a. Karlsburg and Gulafehervar.[3] Sighisoara, which was once Schässburg, and before that Castrum Sex (Fort Six), is now known as the Hungarian Segesvar. These name changes confirm the existence of a multitude of cognitive maps that function as a palimpsest, and they also point to the possibility of multiple, and often competing, identities. Moreover, they indicate the struggle of national languages against each other, the struggle over territory, and the right (not to mention desire) to be heard and remembered.

The struggle over territory is often (if not always) an important element in a given nationalism, and the recent history of South Slavs can serve as a good example. In the late nineteenth century, Serb and Croat national ideologies rested upon dreams of territorial expansion, which were rationalized by the emotionally charged political rhetoric about notions such as »the cradle of civilization« or »the land of our ancestors« that, in turn, needed to be reclaimed »once and for all.« These two national ideologies came into conflict over the question of the territories to »be liberated from foreign rule,« as the South Slavic politicians of the time used to say. The dominant discourse of the national elites of the period, however, was not about *liberation* but rather about *appropriation* of space and its incorporation into either Greater Serbia or Greater Croatia. With this in mind, it would be prudent to revise the canonical view of the so-called »period of national liberation« of the South Slavs and to re-examine historical developments in the region within the framework of expansionist policies of regional elites.

Throughout the territory of the Kingdom of Yugoslavia, as well as that of its socialist successor state (the SFR of Yugoslavia), renaming and appropriating spaces seems to have been standard practice. This is not to say that such a practice was not common in earlier periods. During the Ottoman rule of the region, medieval Varcarevo became Varcar Vakuf, only to be renamed Mrkonjic Grad in 1924 to glorify the old king.[4] The town of Duvno went back and forth, being renamed Tomislavgrad in 1925 and then some decades later once again earning the right to its old name. The post-1945 capital of Montenegro also witnessed several attempts at naming and renaming. What was once the Roman Doclea became Podgorica and then Titovgrad or Titograd and then, again, Podgorica. From 1945 to 1992 the northern Montenegrin town of Berane was called Ivangrad, in celebration of the ideological radicalism of the Montenegrin communist leader, Ivan Milutinovic.

The only names to survive in the region are the ancient names of rivers. The calm waters adorning the western Bosnian landscape are still called the Una. Being among the few relatively stable features of spaces we inhabit and name, rivers have often been used as either a reminder of an old boundary between »us« and »them« or as a modern-day no-man's-land separating former combatants and groups that have not been able to live together any longer due to »irreconcilable differences.« At a time of heightened nationalistic instincts and in spite of the much used rhetoric of inclusion, multiculturalism, and reconciliation – as is still the case throughout the former Yugoslavia –, rivers once again function to divide along the lines of ethnic and national belonging and/or religious affiliation. For hundreds of years, the river Drina was a dividing line between Catholicism and Eastern Orthodoxy, as well as between different ethnic and national groups in the region. In the minds of many Serbian nationalists, the Drina's banks still mark the western boundary of their beloved »ancestral land.« Not long ago, the river Neretva was enforced as a bloody demarcation line between two communities in the city of Mostar. Aside from cutting off lines of communication and disabling physical contact between people living on the opposing riverbanks, the recent destruction of a beautiful medieval arch over the Neretva was also a symbolic cutting off of inter-ethnic and inter-religious ties that for centuries had bound the citizens of Mostar together. Similarly, the KFOR checkpoints along the banks of a small stream of water winding through the city of Kosovska Mitrovica are a potent symbol of the current ethnic and national divisions in the Kosovo region. Nationalistic obsessions with carving out and reinforcing such division lines might be the reason why bridges are intensely hated and attempts made to destroy them whenever possible.

Frightened by the power of space's silence, we try to fill it up with our own names, voices, and groans, and anchor our bodies in it, as if each of us were unconsciously trying to define our own thoughts and ideas about our selves. Like ants, we rush to reach the top of a shaky pyramid of grass and dirt in the hope that our efforts to thicken its walls will provide security and stability for our uncertain identities. While struggling to define our selves by the shape, size, height, and thickness of the walls that surround us, we feed a Janus-faced notion of commonality (tribal, ethnic, national). At least for a short period of time we become space, and space becomes us, but all we are doing is endlessly naming and renaming ourselves and anchoring our notion of belonging and home deep inside an imaginary landscape because we are reluctant to admit that our daily lives revolve around issues of heterogeneity and difference. Naming space, moulding it, and filling its void with signifiers of ethnic, religious, and national belonging are manifestations of human power relations. From the time of the first drawings adorning the walls of the caves at Lascaux to the Space Needle in Seattle, those with power have constructed and reconstructed spaces, and filled them with meaning, thus confirming the dual function of those spaces: as tools of self-identification and as markers of the balance of power in a given society.

It could also be said that in a rather unusual way, and through a set of mnemonic devices, the destruction of space/place serves the same dual functions. This seemingly reversed logic is particularly relevant with regard to the recent events in the former Yugoslavia. Viewed through the short-sighted spectacles of expansionist nationalism, the destruction of space paired with dramatic and violent demographic changes and its subsequent re-naming seemed to serve as preconditions for »A New Beginning« and marked initial stages in the process of recomposing old or constructing new signifiers of identity for both people and space. The notion of »ancestral land« that has to be repossessed from an unwanted *other* drives the nationalistic-minded obsession with renaming and recomposing spaces and identities, and through destruction and violence tries to achieve what is thought to be an ethnically pure Garden of Eden. The post-Dayton division of Bosnia and Herzegovina formalized the existence of an entity known as the Republic of Srpska, which led to 13 towns and cities being instantly renamed or having the ethnic prefix »Srpsko« (Serbian) added to their names. The sections of Sarajevo controlled by the Serbs became Srpsko Sarajevo (Serbian Sarajevo), while the new names of towns such as Srpski Drvar, Srpski Sanski Most, Srpski Mostar, Srpsko Gorazde, and Srpski Kljuc marked the outer edges of the newly created ethnic paradise.[5] The old town of Foca is now called Srbinje and, like other renamed towns, its new name carries a strong Serbian ethnic marker.[6] The town formerly known as Bosanski Novi seems like a different story of renaming. It recently acquired a new set of clothes through its new name: Novi Grad (New Town). While, unlike Srbinje, the new name for

Bosanski Novi carries no obvious ethnic symbols, it does project a strong message about the aforementioned »New Beginning.« In the minds of nationalists, this new arrangement of a space constructed (reconstructed) as it has been, communicates the message that everything is finally in its rightful place, just as it should have been from time immemorial. What has been created is a new landscape, out of which bleeds the lights of tribal fires, a place where the restless night air invokes long gone, but never forgotten, ancestors (real or imagined).

In spite of such invocations, what we see is indeed a new landscape because what once was can never be duplicated. It is a new landscape and a new space because it is the product of our dreams. We dream of satiating our longing for home by re-creating a collage of past lives in a space we remember as once having been »inhabited« by those lives that is now vacant. Such re-creation usually has no real bearing on our past lives; rather, it constructs out of hopes imaginary relationships with space and its silence, and thus possesses many elements of fantasy. Like Callimachus, the famous third-century B.C.E. cataloguer of the ancient Alexandrian Library and the author of *Pinakes*, many of us are also nostalgic, since nostalgia is a longing for an ideal presence that is no longer and might never have been.[7] Such sentiments are not only expressions of loss, but also of displacement. Moreover, they are a love affair with one's memory of past lives and a fantasy that is frozen in time.

Within the growing body of literature that concerns itself with issues of the construction and/or reconstruction of spaces and identities in the Western world, it is possible to distinguish between two main thematic clusters. The first involves the positive promotion of a space understood as a location. Building an image or brand and reconfiguring cities by placing an emphasis on urban tourism has been used in the process of regenerating urban spaces and is in marked contrast to earlier negative urban imagery characterized by decay, favellas, and industrial »graveyards.«[8] The second thematic flow is somewhat cynical in its strong emphasis on the »fake« elements in this process of reconstructing space, where home is a shelter for daydreaming and, we are told, creates »fake« identities and lifeless life. Advocates of this analytical approach point out that the power of capitalism and its ever-growing appetite to exercise social control remains hidden behind the glitzy and multicolor façades of entertainment and fantasy (Las Vegas, for example).[9]

In the Balkans in general and throughout the former Yugoslavia in particular, we are currently witnessing a process of reconstructing spaces and identities that is very similar in character and aim with reconstruction efforts in the Western world. This similarity, however, is somewhat hidden behind layers of colors thought unique to the local Balkan framework. Upon closer examination, it turns out that what we are seeing in the former Yugoslavia is an awkwardly applied collage of different (architectural and cultural) elements/signifiers of the Western model for revitalizing spaces (urban spaces in particular) and reshaping identities. Even though they closely resemble Western models, such architectural and cultural simulacra are indeed new to the former Yugoslavia in terms of content and message, which are intended for both foreign and domestic consumption. Painting it over with local colors might indicate the sudden awakening of the local population to the realities of life in a multiethnic and multicultural space, while the retention of a few signifiers of past lives and (cultural and religious) practices highlights the notion of being well grounded within a given tradition. Striking such a »balance« after the bloody conflict in the region seems to enjoy certain level of acceptance in the West. Even though the daily lives of people throughout the former Yugoslavia might actually function according to very different codes, the representational character of »change« is, nonetheless, viewed in the West as a clear indicator of progress. People wearing Armani suits, Ray Ban sunglasses, and traditional Muslim dress intermingle in the lobbies of new office buildings designed according to Western standards. For many Westerners these are signs that life in the Balkans is getting better and dreams of coexistence are attainable after all. What is sidelined, however, is the fact that dreams seldom come true.

The message intended for domestic consumption is similar: one of hope and progress. There is, however, a slight problem with such a message. The domestic effort to re-create space and reassign meanings is a promise based on a contradiction because it advocates change while claiming continuity. It is a promise of a new beginning or, as Hannah Arendt put it, a »uniquely human way of ordering the future.« It is also a promise of complete reliance on tradition since we always have to remain connected to our past, and yet must pull as far away from it as possible. To remain in touch with our past presupposes constant imaginative effort and a love of memory. Because such love of memory in the former Yugoslavia mutated into an obsession that fed on the blood of »unwanted others,« it seemed most appropriate to re-create spaces marked with blood and scattered bones by Hollywoodizing them: leveling all graves and then erecting over them clean architectural forms which would hide the disturbing contents of the ongoing inner struggle with our memories of recent events. New structures are meant to hide pain and encourage people to try and forget past sufferings. Such hasty external beautification of pain, suffering, and displacement, which still remains largely unpronounceable, disturbs the basic premise of our relationship to space/place: it depersonalizes it.

Our relationship to space/place is inherently personal. It is all about persistent appropriation coupled with our unwillingness to admit or our inability to grasp human ephemerality, and the fact that we are only a percentage in the final reckoning. This relationship consists of two major building blocks that, at the start of a journey, differ somewhat from one another. The difference could be defined as that between *my beloved landscape* and *the landscape of my beloved.* The former signifies notions of space in terms of geography (territory), which is imagined (recognized) as ours and ours alone and has symbolic and strategic value. This is a tale about a tree, its roots and its branches, and a story about Antheus: a conservative myth about homeland which keeps us immobile. It is also a dream of the colors of (our) paradise, of an exquisitely beautiful female who shall, in the end, fall prey to one of the warring parties. The latter is the projection of the idea (ideal) of home, belonging, and the softness of a woman's skin, wrapped in a warm blanket of memories of the good old days. In other words, it is the relentlessly male portrait of an imaginary geography of the soul and an effort to change the nature of time and materialize it so that we could revisit the past much in the same manner we are able to walk through a neighborhood we used to live in. In the end, the differences evaporate and the two notions merge. We find ourselves stranded an ocean away from our Ithaca, away from *our beloved island* on which *our beloved* shall forever remain unreachable. Thus, the island becomes a symbol and a substitute for both: longing for space and for the intimacy of a collective's social poetics. Forever out of reach, the island is *my beloved landscape* and *the landscape of my beloved.* Our relationship with it is a romance, and an often tragic love story, with all the twists and turns of a Hollywood blockbuster: love, affection, sensuality, betrayal, anger, revenge, regrets, blood and tears. It is a tale of picking up scattered remnants of our past selves and putting back together the puzzle of identity in a slightly different way. It is simultaneously an archeology of belonging and the re-composition of being: the traces of our past lives that are hastily sewn together in order to fit as comfortably as possible within the new imagined homeland.

NOTES

1 Milika Pavlovic, »What is it All About,« *Stand* 171 3(2)/3 (3), School of English, Leeds University (2002): 28.

2 Modern-day alpinists often refer to climbing high mountain peaks as »bagging the peak.« The doctor-recommended physical exercise of overcoming natural barriers in order to maintain a healthy lifestyle is, therefore, motivated by a thrill of conquering, and such effort is awarded not only with medals, a leaner body, and a desirable shape, but it is also followed by an overwhelming sense of satisfaction of appropriating.

3 Ken Smith, »Hungarian Quartet,« in *Shed: Selected Poems* (London: Bloodaxe Books, 2002): 264-6.

4 Mrkonjic Grad was named in honor of the Serbian King Petar Karadjordjevic, whose *nom de guerre* during the Herzegovina uprising against the Ottomans (1875-1876) was Petar Mrkonjic.

5 On 16 July 2004, the government of the Republic of Srpska accepted an earlier ruling by the Constitutional Court of Bosnia and Herzegovina, which found the renaming of towns and cities by adding the prefix »Srpski« (o) unconstitutional. Even though the government of the Republic of Srpska went along with the court's decision, its ministers decided not to entirely abandon signifiers of ethnicity and appropriation. They issued a decree renaming towns and cities once again. This time around Srpsko Sarajevo became Sarajevo of the Republic of Srpska, while formerly Srpsko Gorazde was renamed Gorazde of the Republic of Srpska, and so on.

»Vlada RS utvrdila predlog novih imena za gradove,« *Vijesti*, July 16, 2004, *PCNEN*. Available at: http://www.pcnen.com/vijesti/index.php?a+show&vid+1368. Accessed on 17 July 2004.

6 During the same session on July 16, 2004, the government of the Republic of Srpska proposed that the town of *Foca* (*Srbinje*) should have two names: *Foca-Srbinje*. It is important to note that these changes with regard to naming and renaming came as the result of pressure exerted on local political leaders by the UN High Representative in Bosnia and Herzegovina, Mr. Paddy Ashdawn. See endnote 5.

7 »I wept as I remembered how often you and I had tired the sun with talking and sent him down the sky.« Callimachus on nostalgia. See Donald G. Davis and Wayne A. Wiegard (eds.), *Encyclopedia of Library History* (New York: Garland Publishing, Inc., 1994). See also Svetlana Boym, *The Future of Nostalgia* (New York: Basic Books, 2001), »Introduction,« 14-16.

8 Hilje van der Horst, »Multicultural Theming in the Netherlands: Pacifying, Essentializing and Revanchist Effects,« in Susan Ingram and Markus Reisenleitner (eds.), *Placing History: Themed Environments, Urban Consumption and the Public Entertainment Sphere* (Vienna: Turia & Kant, 2003): 175-201.

9 Ibid.

Running on Empty:

Berlin as a Chronotope of Persistence in *Run Lola Run*

SUSAN INGRAM

»Forget Berlin!« That was Lutz Koepnick's exhortation to German Studies scholars making the spatial turn and exploring sites of memory in a special issue of *German Quarterly* in 2001. But can one really forget Berlin? While some cities have persisted as insistently in the collective memory of their nations, something attested to by the work of Victor Burgin among others, fewer have been as indelibly marked by, and have been as resilient in the face of, historical trauma as Berlin, and let's face it, few have had to be: »Since 1871 Berlin has been Germany's capital – a status interpreted in different ways by different regimes, but one nevertheless acknowledged by the Hohenzollern monarchy, the Weimar Republic, the Third Reich, the four Allied occupiers, the two postwar German states, and the unified Germany of the 1990s« (Ladd 3). Only Shanghai springs to mind for comparison, albeit imperfectly as Wolfgang Schivelbusch points out. While the International Zone in Shanghai may have amounted to »nothing other than a European enclave, and [was] in no way the result of a previous conquest« (Schivelbusch 7), the fact remains that Shanghai too, in its modern incarnation, was physically sundered by geopolitical forces beyond its control into zones (or concessions as they were called), it too was considered »liberated« in 1949, and then, much more recently, it too has been massively rebuilt and restructured by the forces of national and transnational capital. It is this latest phase – the traces of historical memory that have persisted in the cityscape of reunified Berlin and been captured on film – that will be my focus here. As Brian Ladd has detailed with great sensitivity in *The Ghosts of Berlin*, the debate following the *Wiedervereinigung* of the German Democratic

Republic and the Federal Republic of Germany on October 3, 1990, on whether and just how much of the federal government to return to Berlin from Bonn,[1] had as one of its many consequences a series of massive architectural projects »intended to restore, reveal, and reinvent both the physical and the symbolic city of Berlin« (Ladd backcover). Now, slightly over a decade since this attempt to bandage over the wound of the Wall and initiate a healing process, *Baustelle* Berlin is itself becoming another figment of the city's past, and Berlin continues to persist, hovering on the brink of yet another crisis, namely, bankruptcy, and dealing with yet another unification, a larger one this time: the aptly namely EU »enlargement.«

The pain and urgency with which the past decade in Berlin has been experienced is reflected in many, if not most, of the artistic works produced in and about it during this time. It is, therefore, all the more extraordinary, at least to me, that what is arguably the best known of these productions, and certainly the one to have made the greatest impact globally, namely Tom Tykwer's 1998 *Lola Rennt* (*Run Lola Run*), is so rarely situated in its historical and geopolitical contexts. Like Walter Ruttmann's 1927 *Berlin: Symphonie der Großstadt* (*Symphony of a City*) and Wim Wender's 1987 *Der Himmel über Berlin* (*Wings of Desire*), Tykwer's film is a chronotopic city-film. It captures, marks, and offers a thesis about a defining moment in the city's history. In the footsteps of Ruttmann's critique of the modern capitalist moment of the 1920s and Wender's melancholic musings about mortality in the face of the division into East and West, Tykwer's is a breathless yet somber reflection on the insurmountable hurdles facing the city in the wake of reunification.

1998 marked a monumental crossroads for Berlin and provided a potent backdrop for a film about a young couple that has twenty minutes to replace the 100,000 German marks the young man has inadvertently left behind on a subway train or face the wrath of his mobster boss. On 23 April 1998, the German Bundestag voted by a great majority for Germany to join the European currency union, in which the new Euro was to be introduced on 1 January 1999, effectively relegating the mark to the dustbin of outdated currencies, while the fall elections that year brought about an equally momentous changing of the guard. Many had expected Helmut Kohl, generally regarded as the driving force behind German unification, to step down before the September election and simply hand over the reigns to Wolfgang Schäuble, his designated heir apparent. He didn't and lost the chancellorship for the first time since 1982. While Gerhard Schroeder's victory was commonly ascribed to widespread disappointment and disillusionment with the way reunification had been (mis)managed, the scandals in which Kohl became enmeshed after the September 1998 election offer a different perspective.

First, irregularities were discovered in an arms-deal, in which the German government sold some tanks to Saudi Arabia.[2] An investigation by the Public Prosecutor's office in Augsburg in 1995 »uncovered the fact that several hundred million marks in bribes had flowed into dubious channels. Investigators [in 1999] allege[d] that at least 1 million marks landed in the accounts of the CDU,« Kohl's party. Deeper digging revealed »striking parallels not only between the [government] bank accounts used, but also the mediators employed in the sale of tanks to the Saudis and the sale of the East German Leuna refinery, including the lucrative Minol petrol station chain, to the French oil company Elf Aquitaine.« It turned out that, in the meantime, the CDU had also received the largest private donation on record, »valued at 2.4 million marks in cheques, plus an interest-free loan of 2.5 million marks [… from] Hamburg real estate dealer Karl Ehlerding« under rather suspicious circumstances:

> In June 1998 the government sold 112,000 apartments belonging to the state-owned railway company to a consortium of firms under the direction of Ehlerding's real estate firm, WCM. Ehlerding's firm won out despite the fact that a Japanese company had made an offer worth around a billion marks more, raising suspicions of a corrupt connection between Ehlerding's donation to the CDU and his firm's lucrative real estate deal with Kohl's government.
>
> In a television interview, Kohl justified the decision to grant the contract to Ehlerding by pointing out that housing privatisations were »always very complicated.« He further asserted that, from the standpoint of the tenants' interests, selling the apartments to the Japanese would have been impossible. In the same interview, Kohl admitted receiving up to two million marks in cash from anonymous donors between 1993 and 1998, which were not listed as donations as prescribed by law. These funds were

said to have subsequently found their way into the official finances of the party.

The next day, the CDU head office contradicted this claim, saying the 2 million cited by Kohl had not been registered at the party treasurer's office. To dispel suspicions of corruption, CDU General Secretary Angela Merkel requested that Kohl, the party's honorary chairman, reveal the names of the donors, but Kohl has refused, saying he gave his word of honour to the donors not to name them.

The sense of scandal brewing around the sets of *Run Lola Run* is neatly captured by the cover story of the 3 July 2000 issue of *Der Spiegel* on »Die Staatsaffäre des Helmut Kohl« (Helmut Kohl's affair of state), which also shows the influence of Tykwer's film: »Helmut Kohl *verspielt* die Reste seiner Reputation, die er eigentlich retten will. Dass in seinem Kanzleramt systematisch Daten gelöscht wurden und Akten verschwanden, macht die Parteispenden- zur Staatsaffäre« (italics added: Helmut Kohl is *gambling away* what's left of the reputation he actually wants to save. That data were systematically erased and files disappeared in his chancellery makes the Party donation affair an affair of state).

In the directorial statement on *Run Lola Run*'s official website, Tom Tykwer claims that the film: »could just as easily be set in Peking, Helsinki or New York, the only thing that would change is the scenery, not the emotional dimension« (http://www.sonypictures.com/classics/runlolarun/runlolarun.html). The purpose of this paper is to debunk that statement. *Run Lola Run* could not have been made anywhere but Berlin. Nor could the same film be made in Berlin today, seven years on. *Run Lola Run* manifests a specific cultural historical imagination, one perhaps not as seductive as the film's video-game aesthetics and destiny-driven philosophy, which have attracted the bulk of the critical attention on the film, but one that, especially now that it no longer specifically obtains, more clearly reveals the pounding rhythm of the city's peculiar dynamism. Tykwer may be right about *Run Lola Run* being a love story, about the way, to quote again from the website, »Love can move mountains, and does. Over and above all the action, the central driving force of this film is romance.« What he neglects to mention, however, and what I will show here by first locating the action and then unpacking the key theme of paternity is that this romance has a noteworthy geopolitical dimension, one which opens the film up to being read chronotopically as being about the very troubled romance of reunification.

Mikhail Bakhtin borrowed the term chronotope from Einstein's theory of relativity and adapted the concept of »the inseparability of space and time« to facilitate exploration of the ways in which these space-time intersections appear in artistic texts (Bakhtin 84-5). While Bakhtin believed »that the chronotope could be used as a medium for appreciating the interrelationships between ›real historical time and space‹ [...] and ›actual historical persons‹ and the expression of these into literary forms« (Jones), he was also aware of the danger of confusing the chronotopically respresented world with the world outside the text (Bakhtin 253). Rather, as scholars of Bakhtin and cinema, such as Robert Stam and Hamid Naficy, have pointed out, the concept of the chronotope draws attention to the specific ways that art mediates historical realities. Chronotopes work »both as a ›unit of analysis‹ for studying text in terms of their representation of spatial and temporal configurations and as an ›optic‹ for analyzing the forces in the culture that produce these configurations« (Naficy 152). They can be, and have been, used to analyze all forms of narrative in text, screen and cyberspace. Examples include Naficy's work on accented cinema in which he traces the specific ways that displacement and emplacement figure in the works of exiled and diasporic filmmakers, Lainie Jones's analysis of the 9/11 attacks on the Twin Towers via linkages between »the real events and fictional events that foreshadowed it« (Jones), Alexandra Leontieva's exploration of relationships between the television series *The X-Files* and its fans, Irina Makoveeva's examination of »The Chronotope of the Maternity Home« in contemporary Russian women's fiction, and Paul Smethurst's identification of a »postmodern chronotope« in regard to the ontology of multiple worlds in cyberspace and contemporary fiction. The following reading of *Run Lola Run* is intended in this spirit of the inseparability of not only time and space, but also fact and fiction.

Run Lola Run opens with a telephone call between Lola and her boyfriend Manni. Where is she? At home, in an apartment on Albrecht Strasse in Mitte, the historical center of Berlin, which was displaced by the building of the Wall and became the site of substantial regentrification after it came down, and which was described in the »Globejotters« section of a Hong Kong newsmagazine as »a hyper-trendy, post-Wall enclave […] which is crowded with cafes, IT companies and museums. Nearly everyone living and working in Mitte comes from elsewhere in Germany [and beyond, one might add]. And all have arrived within the last decade« (Globejotters). Manni, on the other end of the line, is in a telephone booth across town. Specifically he is across the street from a Bolla supermarket on the corner of Osnabrücker Strasse and Tauroggener Strasse, which is a quiet area quite far into what used to be West Berlin, just north of the Spree across from Schloss Charlottenburg. She is in the former East; he is in the former West, and while they both undertake the same journey over the course of the film – from East to West –, Manni's is much shorter. After a quick drug deal on the other side of the Polish border, he remains predominantly in the former West, while Lola spends most of her time in the former East. After exiting the apartment and running past a cemetery, she first visits her father's bank by traversing Friedrichstrasse and the Gendarmmarkt, the area of the city most transformed by the influx of national and transnational capital since reunification. The beautiful red brick bridge she crosses on the way is the Oberbaumbrücke, which connects Kreuzberg with the formerly Eastern district of Friedrichshain. The casino she gambles in, the disguised Kronprinzenpalast, is also in the former East, at the end of Unter den Linden right across from the Palast der Republik, the former East German House of Parliament. It is only in the vicinity of the Bolla supermarket, where Manni is impatiently waiting for her, that Lola enters a site formerly in the West.

When this geographical dimension is factored in, the film's plot, which is iterated three times, becomes the story of someone in the former West calling up someone in the former East and saying »help me, I've screwed up, my gangster-boss is going to kill me, you've got to get me out of this mess,« and it takes three tries for them to finally connect and walk off together hand-in-hand with the gangsters off their backs and 100,000 marks in the bag. However, *Run Lola Run* isn't just a romance – it's a family romance. One of the film's key themes is mistaken paternity. After the opening telephone call, Lola starts spinning around, trying to think of someone with access to 100,000 marks, and she hits upon Papa, who is shown shaking his head as though in response. Indeed, during the first episode, we learn that he is not really her father. When she asks him for the 100,00 marks to save Manni, he becomes angry and tells her that she is a »*Kuckucksei*,« i.e. illegitimate. He does so because he believes that his lover, Jutta, is pregnant with his child; however, in the second episode, a repetition of the first with important differences, we and he learn that Jutta is pregnant by another man because he has neglected her just as he neglected Lola's mother. By the end of the third and final episode, it becomes clear who Lola's real father is: the security guard in Papa's bank, who in the first two episodes lets her into the secured part of the building containing Papa's office and in the third greets her on the building's threshold and then presumably has a heart attack because he is in the ambulance Lola hitches a ride to the Bolla in, and when questioned by the attendant about what she's doing there, she replies, »*ich gehöre zu ihm*« (I'm with him, more literally: I belong to him, I'm his). She then takes his hand in hers, resuscitating him where the ambulance attendant had been unable to do so and returning his heartbeat to normal. The tripartite unfolding of the plot thus suggests the successful resolution of Lola's initial mistake about her paternity.

However, paternity also has a symbolic dimension in the film. That it takes three tries for Lola and Manni to get the money in time and walk off together hand-in-hand is usually interpreted, even by mainstream film critics like Roger Ebert, in terms of chaos theory and the intimately connected causality known as the butterfly effect, according to which a butterfly flapping its wings in Madagascar can cause weather changes around the world. This element of chance in the film works by throwing Lola's split-second timing off in the third episode, which sets off a domino-like chain of events that ends not with either Lola's or Manni's death, as the first and second episodes do, but rather with a deadly collision of father figures. A car accident figures in each of the three episodes (as does a robbery at gun-point and one of Lola's glass-shattering shrieks), but in the first two, one car merely accelerates into another, a mere fender-bender ensues in which no one is hurt, and the gangsters emerge from their car and menacingly approach Papa's friend, Herr Meyer. In the final episode with the Hollywood happy ending, on the other hand, it is a serious crash. Herr Meyer and Papa are killed, while the gangsters once again emerge unscathed.

This final collision I read as reflecting the post-reunification havoc in Berlin's and Germany's economies. The boring, grey suit-clad representatives of the official reunified German financial sector are done in, as is an innocent, environmentally friendly cyclist, while for the funky sunglasses-sporting representatives of an alternative economy based on illegal cross-border drug-running, it's business as usual. It is striking that, unlike the accounts of reunification written for the most part by former West German economists and political scientists, which for the most part take the East to task for the failures of reunification, the film ascribes no blame. The young couple is positioned like the city of Berlin itself: »on the margins of dominant late capitalist time: Lola is unemployed and Manni is a small time crook« (O'Sickey). Neither's prospects are promising, yet Lola is clearly the stronger of the two, the one with the drive, the energy and the answers. When Manni calls her for help, he taunts her that he has landed them in a situation not even *she* can think of a solution for.

The film can thus be seen to rather subtly tap into the emerging groundswell of *Ostalgie*, which erupted with the prominent *Ostalgie* film, *Sonnenallee* (1999, dir. Leander Haußmann) and may have crested with Wolfgang Becker's 2003 *Goodbye, Lenin!* (See also Bach; Berdahl; Nadkarni; and the *Seminar* special issue).[3] Becker clearly spells out his positions for his audiences. There is a piece of graffiti in his previous production, the 1997 *Das Leben ist eine Baustelle (Life is All You Get)*, for which Tom Tykwer co-wrote the screenplay, which reads: »Die Liebe in den Zeiten der Kohl-Ära« (love in the time of Kohl's era, which in the original sounds exactly like the Garcia Marques title). Tywker, on the other hand, is more subtle, gently and symbolically reminding viewers that East Germany was the most prosperous and resourceful Warsaw Pact country and that only in orientalizing Western stereotypes was it imagined to be a place of crime and corruption.[4] Tywker simply encourages viewers to reflect on the kinds of crime and corruption present in the reunified Berlin, very much in the Bakhtinian spirit that »social and political events gain meaning […] and are illuminated […] only insofar as they relate to private fates« (Jones).

The way Chancellor Kohl's Christian Democratic government had been conducting state business in Bonn was just beginning to make the German news when the film was released:

> French and Swiss examining magistrates had already uncovered the fact that the sale [in 1992] of the Leuna refinery and Minol [gas] stations was accompanied by bribes and the payment of dubious commissions of at least 100 million marks.
>
> Referring to Kohl and former President François Mitterrand of France, Loik Le Floch-Prigent, the Elf chief executive at the time of the purchase, told French prosecutors on June 19, 1997, ›I was pushed [into it] by the president of the republic and by the chancellor.‹ In a statement obtained by the French daily *Le Monde*, he continued, ›It was a question of demonstrating that France was present in the process of German reunification and, for Germany, of being able to show off that presence.‹ (http://www.globalpolicy.org/nations/corrupt/kohl7.htm)

Rumors had also begun to circulate that a part of those 100 million marks had found its way into the CDU's coffers. Already on 26 February 1998, *Die Zeit* was posing questions like »Did Bonn illegally subsidize the Leuna refinery?« (http://www.zeit.de/archiv/1998/10/aaa.txt.19980226.xml) and referring to Poland, France and Germany forming a Troika (»Polen, Frankreich und Deutschland bilden eine Troika«). The details reported about how the German arms deal with the Saudis was carried out are also noteworthy. To quote one news report:

> Descriptions of the transactions read like a seedy crime novel. In the parking lot of a Swiss supermarket close to the German border in the summer of 1991, the weapons dealer Karlheinz Schreiber handed over a suitcase stuffed with cash – 1 million in thousand mark bank notes – to the then-CDU treasurer Walther Leisler Kiep and Kohl's confidante Horst Weyrauch.
> (http://www.wsws.org/articles/1999/dec1999/kohl-d30_prn.shtml)

Run Lola Run thus set the stage for its German viewers to think about who the »real« criminals were, about the options available to generation Xers to earn a livelihood of some kind, and also about the other kinds of options the older established generation had quietly and comfortably been accumulating.

The casino sequence in the final episode and all of Lola's high-pitched, glass-shattering, fate-altering screams speak to this desire for agency on the part of the younger generation. However, the gendering of this agency as it is depicted in the film draws attention to the film's not really so happy ending. In the final sequence, after Lola gets out of the ambulance, as Ingeborg Majer O'Sickey relates:

> the film's hyper kinetic energy that the viewer has come to associate with Lola, is drained from her and transferred to Manni. A number of visual cues communicate this message; shots of Lola's background are devoid of any city movement what so ever, whereas shots linked with Manni show a vibrant city life. His swagger as he walks toward her is unmistakable in its message that he has taken command. When he meets up with her, he says ›What's the matter with you? Did you run? Don't worry, everything's O.K.,‹ nearly echoing the words [Papa…] had said to her when she pleaded with him to give her the money (her father had said: ›What is the matter with you? You look terrible!‹). If it were not for Manni's affectionate tone and gestures, one might without hesitation claim that Manni is from that moment on folded into the patriarchal line that Lola's father represents. It is difficult not to read the close-up of the two lovers' intertwined hands as they walk down the street as the film's self-reflexive comment that it understands that it is delivering the couple into a happy ending in the tradition of classical Hollywood cinema's economy of desire. (O'Sickey 131)

The small-time criminal associated with the West steps into the surrogate-father's shoes, and life for the spunky young woman from the East has literally become a crap-shoot à la Baudelaire and Benjamin.

While the film would thus seem to be a rather bleak commentary on the generational shift accompanying reunification, the fate of Lola's real father, the security guard, offers a mitigating viewpoint. He, after all, is the one Lola resuscitates in the final episode, not Manni, who is able to get the money back by himself. The security guard is also the first character to speak to us in the film, during the opening credits. Critics have fastened onto his opening soccer-related comments – »The ball is round. The game lasts 90 minutes. That's a fact. Everything else is theory« – as well as one of the opening epigrams, by the legendary German soccer coach, Sepp Herberger, who took Germany to victory in the 1954 World Cup: »After the game is before the game.« O'Sickey's article on time and desire in *Run Lola Run* is representative in this regard, and she is quite right that the film »practically begs us to think about [it] in terms of its management of time and space. [...] In its most basic sense, the film is a demonstration of Benjamin Franklin's formula ›time is money‹« (123). However, as Bakhtin indicates in »Forms of Time and Chronotope in the Novel« and as I hope to have shown here, one should not neglect the spatial dimension, but rather should note how »space becomes charged and responsive to the movements of time, plot and history« (Bakhtin 84-5).

One final point about the chronotopic quality of *Run Lola Run* with which I would like to conclude is that it provides an intriguing link to the 1997 Hong Kong triad flick, *Yi ge zi tou de dan sheng* (*Too Many Ways to be Number One*, dir. Ka-Fai Wai).[5] While *Too Many Ways to be Number One* is often likened to *Pulp Fiction*, *Reservoir Dogs* and *Sliding Door*,[6] its resemblance to *Run Lola Run* is palpable. Besides the striking cinematography, in particular in the red-drenched bedroom scenes, *Too Many Ways to be Number One* notably features a tripartite structure in which the protagonist's three possible fates are played out. In the first episode, he goes with some fellow gang members to the mainland and is killed; in the second he goes with a friend to Taiwan, is badly beaten by local gangsters and ends up in a wheelchair; in the third and final, open-ended episode, he remains in Hong Kong, and the viewer is left to presume that, like Lola, he has learned the lessons of the previous two episodes and will live reasonably happily ever after despite the changing geopolitical realities of their increasingly globalizing worlds. Given that the film was released in the year in which Hong Kong was handed over to China, the significance of the three settings seems even more straightforward than the subtly gendered relation between East and West in *Run Lola Run*.

The first of Tykwer's film's two epigrams, an old bromide[7] from T.S. Eliot's »Little Gidding,« which are the first words we see on the screen during the opening credits, also supports a chronotopic reading of the film: »We shall not cease from exploration [...] and the end of all our exploring will be to arrive where we started [...] and know the place for the first time.« Now that this paper has arrived at the end of its explorations, I hope we have also arrived at a sense of the pumping heart in Berlin's imaginary and how it manages, and continues, to tick along.

NOTES

1 »Mit einer knappen Mehrheit von 338 zu 320 Stimmen fällt nach hitziger Debatte am 20. Juni 1991 die Entscheidung für Berlin. [...] 12. Oktober 1993 beschließt die Bundesregierung, bis zum Jahre 2000 nach Berlin umzuziehen. Der Umzug der einzelnen Ministerien soll nach und nach erfolgen. [...] Zwischen Bonn und Berlin soll es jedoch eine ›faire Arbeitsteilung‹ geben. Nach dem sogenannten Kombinationsmodell vom 11. Dezember 1991 sollen das Auswärtige Amt sowie die Ministerien des Innern, der Justiz, der Finanzen, der Wirtschaft, für Arbeit und Sozialordnung, für Familie, Senioren, Frauen und Jugend, für Verkehr, für Raumordnung, Bauwesen und Städtebau nach Berlin umziehen. Die übrigen Ministerien bleiben in Bonn. Der Bundesrat verlegt seinen Sitz nach Berlin, obwohl er sich zunächst am 5. Juli 1991 mit 38 von 68 Stimmen unter Vorbehalt für Bonn entschieden hat. Am 3. Juni 1992 legt die Bundesregierung eine Gesamtkonzeption zum Berlin-Umzug vor, und am 25. August 1992 schließt sie mit dem Berliner Senat und der Regierung von Brandenburg den Hauptstadtvertrag ab.

Am 12. Oktober 1993 beschließt die Bundesregierung, bis zum Jahre 2000 nach Berlin umzuziehen. Der Umzug der einzelnen Ministerien soll nach und nach erfolgen. Bundespräsident Richard von Weizsäcker verlegt bereits am 31. Januar 1994 seinen ersten Amtssitz nach Berlin. Am 10. März 1994 verabschiedet der Bundestag das Berlin/Bonn-Gesetz, das die rechtlichen Voraussetzungen zur Umsetzung des Berlin-Beschlusses regelt. Bundeseinrichtungen wie der Bundesrechnungshof, das Bundesversicherungsamt etc. gehen nach Berlin. Im Gegenzug sagt der Bund zu, andere Institutionen, auch internationale, nach Bonn zu ziehen. Am 1. Januar 1995 tritt der Bonn-Vertrag in Kraft, nach dem der Region Bonn vom Bund Ausgleichsmaßnahmen in Milliardenhöhe für den Verlust des Parlaments- und Regierungssitzes bis zum Jahre 2004 zugebilligt werden. Berlin erhält ebenfalls gemäß dem Hauptstadtfinanzierungsvertrag zwischen dem Bund und dem Land Berlin finanzielle Unterstützung bis Ende 2004. (http://www.dhm.de/lemo/html/WegeInDieGegenwart/FolgenDerDeutschenEinheit/hauptstadtdebatte.html). Accessed on 15 May 2005.

2 The unidentified quotes in this section are all from Rippert.

3 As Krisztina Fehérváry notes, »Recent literature on post-socialist consumption, particularly on eastern Germany, has developed around the phenomenon of ›nostalgia,‹ that is, the revival and revaluing of ›socialist era‹ consumer products and activities. Most of this

literature correctly understands these practices not as a desire to bring back the socialist past, but as an intervention in dominant western discourses condemning socialism and its artefacts, devaluing the socialist experience and a process of reclaiming personal pasts (see especially Berdahl 1999; and Bach 2002 for Germany and Nadkarni n.d. for Hungary; also the film »Remembering Lenin« [sic])« (Fehérváry 15, fn5). Fehérváry's intervention in the debate is to remind scholars interested in the region and the topic »that in the main, attitudes towards socialist material culture and the politics it symbolized were overwhelmingly negative. Only select items from the past are revived (something worth analyzing), and, especially in Hungary, only by certain segments of the population. As Nadkarni recounts, hip urban collectors of socialist furniture to decorate a theme-pub were rebuffed when they offered to pay people for this kind of furniture. Accepting payment would have not only contradicted the visceral dislike people had developed for this style, but would have questioned their participation in a community of ›taste‹ which devalued these objects and the material culture they were a part of« (15). For information on touring sites of *Ostalgie* in Berlin, see: http://www.berlin-tourist-information.de/english/unterwegs/e_uw_berlinprogramm_ostalgie.html.

4 This imagery clearly parallels the way everyday life in socialist Poland is represented in *Avalon*, which Markus Reisenleitner discusses in his contribution to this volume, and also is related to the stakes involved in the nostalgic renaming of places in the Balkans that Srdja Pavlovic explores. Thinking these connections together with the theme of the »Floodgates« workshop and the technological expertise of the participants (Simon Woods and Markus Reisenleitner in particular) brings out how useful videogames can be as a metaphor for the *Ostalgie* phenomenon. If one approaches the world economy as a videogame, then *Ostalgie* can be understood as a level achieved once a certain, necessarily uneven influx of capital into previously socialist states has been reached, which, as Reisenleitner's contribution spells out, makes »Class Real« a commentary on having the appropriate weapons and attire to navigate »Class Capital.«

5 I would like to thank Jeff Cheung Gi Hang for not only introducing me to but also lending me his own VCD of this film.

6 These comparisons are primarily to be found on websites dedicated to Hong Kong film. See: http://www.cityonfire.com/hkfilms/st/toomanyways.html; http://chinesecinemas.org/capsules.html#too; http://www.asia.cinedie.com/too_many_ways.htm; http://www.moria.co.nz/fantasy/toomanyways.htm.

7 My thanks to Michael Saffle for this wonderful turn of phrase.

REFERENCES

Bach, Jonathan. »The Taste Remains: Consumption (N)Ostalgia, and the Production of East Germany.« *Public Culture* 14.3 (2002): 545-56.

Bakhtin, M.M. »Forms of Time and Chronotope in the Novel.« Trans. Caryl Emerson and Michael Holquist. *The Dialogic Imagination: Four Essays*. Austin Texas: University of Texas Press, 1981. 84-258.

Berdahl, Daphne. »›(N)Ostalgie‹ for the Present: Memory, Longing, and East German Things.« *Ethnos* 64.2 (1999): 192-211.

Burgin, Victor. *Some Cities*. Berkeley: University of California Press, 1996.

Ebert, Roger. Review of *Run Lola Run*. 2 July 1999. http://rogerebert.suntimes.com/apps/pbcs.dll/article?AID=/19990702/REVIEWS/907020304/1023. Accessed 1 August 2004.

»Globejotters.« *South China Morning Post, Post Magazine* (01.09.2002): 46.

Jones, Lainie. »Life Imitates Art: The Chronotope of the Twin Towers in Fact and Fiction.« *Australian Humanities Review*. May 2003. Available: http://www.lib.latrobe.edu.au/AHR/archive/Issue-May-2003/jones.html. Accessed 1 August 2004.

Koepnick, Lutz. »Forget Berlin.« *German Quarterly* 74.4 (Fall 2001): 343-54.

Ladd, Brian. *The Ghosts of Berlin: Confronting German History in the Urban Landscape*. Chicago and London: University of Chicago Press, 1997.

Leontieva, A. »Bakhtin's Definition of the Adventure Chronotope and the Phenomenon of the X-Files Fan Fiction.« Available: http://www.hf.uib.no/hfolk/henal/fanfic.html. Accessed 1 August 2004.

Makoveeva, Irina. »The Chronotope of the Maternity Home.« Available: http://www.pitt.edu/AFShome/s/l/slavic/public/html/sisc/SISC4/makoveeva.pdf. Accessed 1 August 2004.

Nadkarni, Maya. »›Csak a Szépre Emlékszem‹ [I Only Remember the Beautiful]: Consuming Nostalgia in Post-Socialist Hungary.« *Soyuz Conference*. Ann Arbor, Michigan, February 23, 2002.

Naficy, Hamid. *An Accented Cinema: Exilic and Diasporic Filmmaking*. Princeton and Oxford: Princeton University Press, 2001.

O'Sickey, Ingeborg Majer. »Whatever Lola Wants, Lola Gets (or Does She?): Time and Desire in Tom Tykwer's *Run Lola Run*.« *Quarterly Review of Film and Video* 19 (2002): 123-31.

Rippert, Ulrich. »Former Chancellor Helmut Kohl Faces Criminal Investigation: What Lies Behind the German Christian Democrats' Financial Scandal? World Socialist Web Site.« 30 December 1999. Available: http://www.wsws.org/articles/1999/dec1999/kohl-d30.shtml. Accessed on 15 May 2005.

Schivelbusch, Wolfgang. *In a Cold Crater: Cultural and Intellectual Life in Berlin, 1945-1948.* Trans. Kelly Barry. Berkeley, Los Angeles, London: University of California Press, 1998.

Seminar: A Journal of Germanic Studies. Special Theme Issue »Beyond Ostalgie: East and West German Identity in Contermporary German Culture.« Guest Editors David Clarke and William Niven. 40.3 (September 2004).

Smethurst, Paul. *The Postmodern Chronotope. Reading Space and Time in Contemporary Fiction.* Amsterdam/Atlanta, GA: Rodopoi, 2000.

Stam, Robert. *Subversive Pleasures: Bakhtin, Cultural Criticism, and Film.* Baltimore: Johns Hopkins University Press, 1989.

L'viv *über alles*, an Eden for Intellectuals

ANDRIY ZAYARNYUK

I: INTRODUCTION: A PLACE TO LIVE IMAGINED AS A SPACE TO WRITE

In recent years, Ukraine has become one of the sites in which intellectuals have been consciously experimenting with the construction of identities. Well aware of the role intellectuals have played in all kinds of modern projects, envisioning and then engineering collectives bound not only by formal institutions and organizations but also by the foremost loyalty to an imagined community, these postmodern intellectuals have been trying to do something similar while maintaining a certain distance between themselves and the communities they imagine.[1] In doing so they engage self-consciously with the blurring of boundaries between fiction and academic writing, trying to combine scholarly activity directed towards the academic community with public appeals and texts that can capture the imagination of the populace.

While such a combination of roles can hardly be seen as something new,[2] the technologies these intellectuals are using, the loci to which their energies are applied and, most importantly, their view of what they are doing reveal much about both their concrete socio-cultural context and the discourses created in them and about them. In the past I have attempted to analyze one tendency in these intellectual projects: the construction of an alternative identity to the one associated with the current Ukrainian nation-state, namely, the issue of a separate Galician identity.[3] But the power of intellectual projects and their popular appeal at least partly depends on their flexibility and ability to react to changing circumstances, to keep up with the changing social, political and cultural contexts in which they appear. I believe that since 2000 some important changes have been taking place among the »Galician« intellectuals in question. One of the more important and most openly manifested shifts has been in the site of the projected identity from a »historically« formed territory (with implied loyalty to the bounded space and its potential for sovereignty) to the city as an allegedly more inclusive space detached from troubled nationalist pasts and clichés.

While Andrii Sadovyi, a candidate for L'viv's mayor, could maintain in December 1999 that »we are interested in everything, but about L'viv we talk very little,«[4] this is obviously no longer the case. The production of fiction and memoirs about L'viv in Ukrainian has reached such an extent that one can speak of the conscious creation of a particular Ukrainian »L'viviana.«[5] *Gazeta Lwowska /Lemberger Zeitung*, one of the oldest newspapers in L'viv was resurrected in 2003 in a new Ukrainian version as *L'viv'ska Hazeta*. Of foremost interest for this article is the fact that three periodicals for and by intellectuals published special issues on L'viv almost simultaneously. The first one was in the most dynamic Ukrainian book-review magazine *Krytyka*, the projected audience of which is primarily a widely defined intellectual community in Ukraine.[6] The second to join was the *Harvard Ukrainian Studies* (*HUS*) targeting the English-proficient global academic Ukrainian studies community.[7] The last one was *Ji* magazine, which targets Western Ukrainian intellectuals in particular, characterizes itself as an »independent Cultural Studies periodical,« and represents the non-governmental organization of the same name, whose declared goal is »research on the problems of Cultural Studies, Political Science, and Philosophy. The main directions of the research are the study of interethnic relationships, civilizational cleavages and administrative borders, the formation of a European identity and questions of European integration and contemporary political discourse.«[8] Finally, outside of the purely Ukrainian and in a »post-Habsburg« context, interest in L'viv is reflected in a special forum, »A City of Many Names: Lemberg/Lwów/L'viv/L'vov – Nationalizing in an Urban Context« published in 2003 as a volume of the *Austrian History Yearbook*.[9]

The two Ukrainian discourses in question, that of Galicia and that of L'viv, have always been closely intertwined. That the issue of the journal *Krytyka* set out to deal with discourses about L'viv is evident from its title, »Passions for L'viv,« which was borrowed from one of its articles. Nonetheless, the analysis itself oscillated between the two discourses and two projects – those about L'viv and those about Galicia; many authors saw the former as an integral part of the latter and never tried to make a clear-cut distinction between them.[10] The journal's editors believed that at the time the issue was published L'viv was the only city in Ukraine where the discourse about local specificity, tradition and history had reached a certain critical mass demanding particular attention.[11]

It seems that the »specificity« analyzed in *Krytyka*'s articles comes largely from the statements of intellectuals and not from local narratives or even the city's ethnography. It is not that the particularity of context has somehow been transfigured into intellectual projects, rather on the contrary: selective representations of the city's past, present and future, created by intellectuals, add to their vision of the city's true character. I would argue that these intellectuals are the true *genius loci* of L'viv,[12] actively constructing the object they like to talk about so much. In doing so they self-consciously pose as intellectuals, claiming authority on the grounds of possessing certain expertise based on their scholarly knowledge of the place in question and of the world. Their projects have various dimensions, and the actual reform of the city is not the last issue on their agendas.[13] They operate networks and seek resources in the nexus created by local academic institutions and governmental agencies, local entrepreneurs and politicians, and last but not least Western academia and foundations. (One of the rationales behind the analyzed project no doubt is that in the contemporary global academic and cultural market, the city is a commodity more liquid – more easily converted into money form – than territorial separatism with clear nationalist overtones.) The goal of this article is not to analyze the personnel, networks or actual material operations of this intellectual project. This article's task is to figure out how the city in question, L'viv, is being imagined, what concepts are employed, and what the implications of such a vision of urbanity and society are. Partly based on participant observation but largely on critical reading of these intellectuals' texts, I would like to explore the tensions between their postmodern awareness of the ability to »rewrite« social reality, references to that reality that are supposed to add credibility to their texts, and the normative visions to which these projects can ultimately be reduced.

II: MYTH

One of the keywords in the discourse about L'viv is »myth.« The usage of this word by Ukrainian intellectuals has become so universal that the absence of debate over it is puzzling. Yaroslav Hrytsak was the first to notice the ubiquity of the word »myth« in Western Ukrainian intellectual discourse and the remarkable transformation in how it was perceived during the last decade.[14] Hrytsak might be correct to trace it back to the translation into Ukrainian of George Grabowicz's *The Poet as Mythmaker*.[15] This structuralist analysis of Shevchenko's historical discourse, which broke it down into some basic concepts and oppositions, was at first found absolutely repugnant by a nationalist-minded public who, without delving into the subtleties of Grabowicz's analysis, were put off by the very idea that someone had dared to treat the basic truths of the Ukrainian national creed as mere »myths.« Even so, with time, and perhaps, in conjunction with some developments in academia – with the younger generation of scholars becoming blatantly constructivist – some crude version of social constructivism became commonsensical wisdom among those engaged with and exposed to the public debates about societal transformation in Western Ukraine, the choices to be made and identities to be shaped.[16]

Ironically, the only attempt thus far to analyze the myths which are being actively written about L'viv has been done by the same George Grabowicz who contributed to the ubiquity of the term »myth.« In »Mythologizing Lviv/Lwów: Echoes of Presence and Absence,«[17] he analyzes mythoi of nineteenth-century Polish and Ukrainian narratives without paying much attention to the intentions of those narrating them. For Grabowicz the real distinction between his and others' usages of the term »myth,« in which it turns into »no more than theme or topos,« is its ability to resonate with the context:

> In our context, an adequate use of myth, its cultural (not publicist and political) applicability, depends on breadth and collective resonance, as well as on narrative scope, on the necessary presence of variants, and above all on symbolic and psychological roots, on the articulation of profound truths and values, and indeed traumas and fears, that go to the very heart of a society and its perception of itself.[18]

Such a contextualized meaning of myth comes from Bronislaw Malinowski and his anthropological studies and can hardly be disputed. However, Grabowicz's usage of the term avoids another aspect of myth and of the context in which the designation is used, one that is constantly emphasized by Malinowski: the awareness that myth-makers and myth-tellers have of the specificity of their narratives. These are people who designate with this word not any but a particular kind of story, an especially important one, and at the same time one cardinally different from their daily experiences. In Malinowski's words:

> Now this line of demarcation between the world of myth and that of actual reality – the simple difference that in the former things happen which never occur nowadays – is undoubtedly felt and realised by the natives, though they themselves could not put it into words.[19]

The awareness of this demarcation between myth and reality combined with the ability to articulate it figures prominently in the work creating the contemporary L'viv myth. The word »myth« is used not only by literati and academics but also by publicists, amateur writers and prominent public figures. In the words of one of the amateurs writing L'viv's local history, anything from laws to the internet can fall into the category of »myth,« and myth itself is just another dimension of life, its beautified and narrated »superstructure.«[20] These intellectuals see myth as the »foundation for society and values,« to paraphrase the title of one of the innumerable books on myth.[21] Moreover, they see and style themselves as the creators of this myth.

The different emphases in the usage of the term myth came to the fore during a discussion with Yaroslav Hrytsak on contemporary discussions around L'viv published in *Ji*. While Anton Borkovs'kyi, a proponent and active participant in the intellectual project of mythologizing L'viv, stated that »Myths as such are formed consciously, very often consciously. [... L]et's not forget the extent to which, in general, the whole ideological struggle is mythological,« Hrytsak and Grabowicz preferred to maintain analytical distance from the project and emphasized the reception and resonance these consciously formulated narratives have to achieve.[22]

On the one hand, the fascination with myth would seem to be a much desired development, departing as it does from the notions of real identities and real cultural boundaries with which much Soviet and anti-Soviet identity formation was connected. On the other hand, much about it makes me feel uncomfortable. To start with, it seems that too many processes which form different historical contexts are treated as myths while other concepts that could be used in these cases – ideology, for example – are never mentioned. The ubiquity of the myth and derivatives of it is especially striking in the articles trying to analyze these »myths.« Let's take as an example Iurii Prokhas'ko's article »Kakania or Cekania (on the ›Mission‹ of Galicia).« Prokhas'ko talks about »the resurrection of old and the birth of new Galician myths« and defines the process as »Galician mythmaking.« The Galician myth is said to be based on the Austrian myth of »Austria Felix.« Then he turns to »Habsburg mythology,« mentions new »mythologems« created after the Second World War, and ends with the importation of the »postmodern Galician myth.«[23] While the myth's origin and genealogy are in this case the focus of attention, in many other cases myth is used as a heuristic concept with the only justification for its usage being the popularity of the term. In Matyukhina's book on everyday life in post-World War II L'viv, there is a whole chapter on the Polish myth in Soviet L'viv, which deals mostly not with narratives about Polish times but with those about the very concrete impact on L'viv of the very concrete Polish People's Republic and its cultural processes.[24] The power of this myth talk is such that it finds inroads into even the texts of those who have serious reservations about the term and its usages.[25]

By using the term myth, contemporary Western Ukrainian intellectuals try to distance themselves from what is stated in the narratives they designate in this way and often help to spread. They create their own pictures of the city or their own versions of history without claiming to reconstruct exactly how it is, was or has been, and in doing so they do not forewarn readers about this aspect of their narratives. They do not say that their representations are not based on documentary evidence; they do not draw an explicit line between facts on which some scholarly consensus has been established, those simply gleaned from sources without any kind of critical analysis and pure fictional invention. Moreover, in many cases the rhetoric of the text deliberately aims to create not only a »reality-effect« but also a »truth-effect,« claiming the authoritativeness usually associated with scholarly work.[26]

Although some visions of this alternative mythologizing are very disturbing,[27] these experiments would be entirely legitimate if the issue was limited to the challenging of already established cultural narratives, their demystification and subversion. But this tendency has a different set of implications. Not only is the presence of myth made explicit but the need for myth is also postulated. It is said that the populace cannot do without myth. While older generations of intellectuals seemed to believe that myths needed to be dismantled, these intellectuals see their task not so much as the dismantling of old myths as the replacing of them with better ones. The implication is that new myths will create new identities, which the populace will embrace.

This is not just a Ukrainian development and seems to be connected with developments in much of Central-Eastern Europe and with the urban history of this area in particular. Theoretical elaboration for these kinds of projects has been provided not by Ukrainian intellectuals but by »Western« scholars. In a recent volume on urban history and city identity, Blair A. Ruble has postulated the need for »new sorts of identities and modes of thought that must arise for a democratic transition to become complete: they do so through urban and community myth and history.«[28] He distinguishes between the »physical city,« which is »repossessed by one group or other through selective policies of public and private restoration,« and the »metaphysical city,« which »is restructured by a selective retelling of history through tours, guidebooks, textbooks, films and Internet sites.«[29] For the construction of this latter city, urban history has to fulfill »the arduous mission of identifying a *civitas* that can be sufficiently wide as to embrace all of the varieties that make cities both urban and urbane. Otherwise, disparate and conflicting groups and individuals live dangerously apart even as they live snugly together.«[30]

This connection between representations and the remolding of people's identity is also very much present in the Ukrainian case. The writing and dissemination of myth is done not only for the reading public's enjoyment but also to alter the identity of individuals and public spaces.[31] Intellectuals adhering to social constructivism appear to be more ready to engage in the process of constructing new all-inclusive narratives than in deconstructing the old ones. Instead of critical analysis we encounter the creation of a new coherent body of representation that is supposed to win the minds of the populace.

There seems to be a basic distinction underlying this myth-making project: one between lay and true knowledge. In this kind of thinking there are those who cannot do without myths and those who can create these myths. There are »simple people« who need myths, and the myths that would be good for them should be created by intellectuals. Such an attitude is problematic not only on ethical grounds and prone to charges of being against intellectual integrity and honesty, but it also neglects and discards popular imagination and popular agency. Believing that some kind of »myth-making« is an integral part of everyday life and people's experience, some intellectuals see only their »myths« as those worth accepting and retelling.

The distinction made by Katherine Verdery between »procedural« and »teleological« intellectuals is useful here. Those creating L'viv's new myth fit very well into her description of the »teleological elite«: »For such people, the conduct of politics and intellectual life, as well as their place in these, was defined first and foremost in terms of the pursuit and defense of certain *values*, rather than the mastery and institutionalization of certain *procedures*.«[32] They share this tendency very much with the Communist regimes' intellectual elites that they replaced. Despite the fact that they articulate very different values and orientations, they articulate them in the same mode, having »certain kinds of symbolic capital to prevail over other kinds as political resources: specifically, moral capital – a capital rooted in defining certain values as correct and upholding them.«[33] A similar tendency among intellectuals was analyzed in the Chinese context and identified as elitism and a wariness of democracy as associated with masses.[34] In our case intellectuals' values seep into their representations of the city intended for public consumption and usage.

The concern with representations of the past and with longings for places the way they were and the way they should be is not limited to Eastern-Central Europe but is a much larger phenomenon. Svetlana Boym discusses the construction of representations that come in the guise of nostalgia and tries to draw an evaluative line between »restorative« nostalgia, which tries to fix a snapshot of the past, and »reflective« nostalgia, which acknowledges flux and irrevocability and serves to foster critical encounters with the Self as well as with various collective bodies.[35] In the case of L'viv and of Central-Eastern Europe in general, »myth« and »nostalgia« have a wider currency perhaps because of the greater temporal distance between the images of the past brought in and the everyday experience of those for whom these »snapshots« are being created, but also, I believe, because this emphasizes the project's orientation towards future reform and its potential power to change the way things really are. Significantly, the »myth« does not claim to »restore« something that once had been. Rather it acknowledges the creation of a new myth and thus, at least in appearance, shares certain features with Boym's »reflective« nostalgia. Despite appearances, this new myth seems susceptible to falling into the traps of exclusivist differentiation usually associated with narratives of national identity. In the attempt to replace older representations with new ones and thus bridge some of the cleavages identified as more important than others, the intellectuals in question ignore and sometime actually draw on others, no less important and potentially just as dangerous.

III. URBANIA VERSUS RURALIA

The first distinction constitutive of the new myth of L'viv is one between city and village. This is the well-known distinction in the social sciences between *Gemeinschaft* and *Gesellschaft*, which turns on the opposition between modern, individualistic, anonymous, dynamic, and progressive society, on the one hand, and traditional, collective, face-to-face, slow and stagnant community, on the other. While in the social sciences the distinction between these two appears as an ideal-type approximation, in the current Ukrainian discussion it is constantly emphasized that we are dealing with the very real distinction between a very concrete and named city and unnamed but no less concrete villages.

In his programmatic article »L'viv. Sine qua non,« Taras Vozniak states that the city is not a village and has not develop from a village: »the city was founded on a different law and different chronotopos.« It has its own urban time and its own space which has been adapted for humans. The punctiformity of the city is juxtaposed to the stretched countryside.[36] In the case of L'viv this leads to the »*contradiction between the city as a ›machine for the acceleration of civilization‹ and traditionally rustic Ukrainian society* that slows down the speed of this machine. The aggregation of rustics does not mean the creation of a dynamic city. Although, of course, the city needs an influx of populace, it has to flow precisely into an urban social structure, and not erode it.«[37] An academic monograph articulates similar ideas:

> Old European cities throughout the centuries possessed in general a stable population. Despite the noticeable influx of newcomers, first of all from the village, a kind of a core of inhabitants had been living there for several generations. Precisely this stratum preserves and reproduces the cultural characteristics of a given city, especially features of its specificity. The character of life, its unwritten history – everything that we conditionally designate as the spirit of a city, its *genius loci*.[38]

It seems that some basics of historical knowledge need to be reiterated. The modern city was the creation of the industrial revolution – in the nineteenth century »old towns grew at rapid rates, and new ones grew even faster. [...] The rise of the city did not stem from an internally generated population surplus.«[39] Its social structure had as little to do with the social structure of the old cities as with the social structure of the countryside. The same great changes that created modern cities also created the modern countryside and changed the lives of the rural population dramatically.[40]

Vozniak also gives the example of »pathologies« into which cities may from time to time degenerate: Paris and Madrid, whose social structure in the eighteenth century was allegedly »deformed by big aristocrats«; Berlin under Friedrich II, and Naples in the seventeenth century, when »gay and dirty Naples was swarmed with absolute paupers, who constituted one third of its population and who were connected with the state and city by nothing but the gallows, as was testified by contemporaries.«[41] These examples are used to diagnose the pathology of contemporary L'viv and identify those slowing down the city's machinery and spoiling the city's civilizational glamor – the »newly converted city-dwellers« with their rural »mental rudiments« from the areas built in Soviet times.[42]

In this case we find an extreme version of discredited modernization theory. Defending Matyukhina's idea that newcomers from the Galician villages were the most stubborn resisters of modernization in Soviet L'viv, Ihor Chornovol declares that »the ability of the peasantry (whether Irish, or Polish, or Russian, or Ukrainian, or French) to resist modernization processes is common knowledge.«[43] Of course, if by modernization one means policies by which land held by peasants was expropriated, the disbandment of peasant communities, or Soviet collectivization, then this kind of modernization was indeed resisted by peasants. On the other hand, we know that in some cases peasants actively participated in the building of »modernity.«[44] In the case of Galician peasants, it seems that their enmity towards the establishment of schools, reading clubs, cooperatives, credit unions and numerous other voluntary organizations in the nineteenth and twentieth century was no greater than that of the urban population.

The quintessence of the rural »problem« and problematic rural attitude for Ukrainian intellectuals is Sykhiv, the new and most densely populated city district in southern L'viv built in the 1980s and 1990s. It is mentioned in this sense by Vozniak[45] and characterized as »anti-L'viv, frozen, which extinguished whatever European was left in the city«[46] by another author in the *Ji* issue, the Russian-speaking Grigorii Komskii. In his essay Komskii describes his arrival in L'viv as a child in 1954 beguiled by immersion into the material civilization of Western Europe – doorknobs, floors, furniture, but also some survivors from the old city, whom we characteristically find in his text in the position of servants and housekeepers. This was the L'viv he loved, but then:

In the mid-1970s completely new people came. They came from the surrounding villages to receive city registration. They hated city dwellers with the class hatred of *Lumpen*. Tenants lived on the upper stories while our heroes had to huddle in semi-basements. Tenants had their own property and opaque lifestyle – which was irritating. All the time they were going somewhere, were leaving, coming back, changing outfits and disappearing again. Tenants were urbanites and unwilling to share the poorly articulated and frequently aggressive philosophical discourses of the neophytes. Between them there was a difficult-to-bridge division line between notions about modes of life. An urbanite would defend the inviolability of private life while a former peasant was used to knowing everything about his neighbors, the lack of knowledge or even worse, mystery, were for him unbearable.[47]

Another Ukrainian author, Il'ko Lemko, similarly describes the »rusticalization« of L'viv, the »invasion from the neighboring villages,« which led to the »final loss of its urban culture.«[48]

Later on Komskii moves on to differences in consumption patterns. »Urbanites« would spend money on barbers, coffeehouses, studying, and clothing while »villagers,« who could not afford such things, projected their hatred of the »tenants« onto the estates in which they lived. In this latter part it becomes clear that the bitter experience of the author's youth has little to do with the characteristics of village life but rather with tensions based on inequality of social position and reinforced by ethnic difference. The life of »urbanites« he describes resembles the ideal life of a Western bourgeois and could be afforded only by those with a privileged social position, greater buying power and access to services and privileges.

The fact that much of the distinction often perceived in L'viv and represented as between urban and rural can be attributed to social differentiation was perceptively noticed by Maciej Janowski:

When in the fall of 1994 my wife and I were walking around downtown L'viv sometimes we had the impression that we were surrounded (as it was expressed by my wife) by large theatrical sets, a city that lacked inhabitants. Perhaps, the season was a bit guilty of that (November is far from the best month for travel), but still both of us could not get rid of the impression that crowds of tired, poor and (although this is difficult to explain rationally) rural people are moving around the city. Exactly the same feeling always accompanied me (at least before 1989) in the former German territories that were annexed to Poland after the war and in the times of communist propaganda deceitfully named ›recovered‹ […] Nothing strange – was my rationalization – there is a crisis in Ukraine, L'viv's inhabitants are poor, most of them come to the city from elsewhere and no wonder they create the impression of being ›maladjusted‹ to the pompous ancient architecture among which they have to live. How surprised I was when in an old tourist guide I noticed in a picture that a crowd in L'viv in 1916 was also predominantly rural: here are *baba*s in head scarves, sitting near the fountain on Rynok Square, there is a basket (with eggs?) nearby; and here is a beggar in front of the Boims' Chapel – just exactly like one sitting there now. Then, perhaps, the assumption about inhabitants being ›maladjusted‹ to their own city was biased? For sure, it is mistaken to assume that L'vivites in the past were as dignified, rich and elegant as the buildings in which they lived and worked. Such people could constitute only a negligible minority.[49]

But let us return to Komskii. He believes that in the 1970s:

> Overwhelming population growth catastrophically reduced the percentage of native urbanites, the mechanism of interpersonal relationships – the so-called urban environment – was broken. L'viv had swollen on account of agents of a mentality foreign to it. Bolsheviks had almost managed to erase the distinction between city and village in favor of the latter eliminating the original spirit of the former and expelling from L'viv its *genius loci*.[50]

The irony is that this statement comes from a person who moved to L'viv in 1954, precisely after the most dramatic population change had already happened and the majority of L'viv's pre-war population (should we say »native«?) had left the city. Most Russian-speaking authors' representations of L'viv and the Ukrainian part of the city's population are informed by national conflict fed by memories of the guerilla warfare fought in the second half of the 1940s in the city with no lesser intensity than in the countryside.[51]

In the *HUS* collection Bohdan Tscherkes characterizes the attitudes of post-war newcomers in the following way: »For most postwar settlers, the city's historical structures held little significance or meaning for their community, and some even viewed the unfamiliar environment with actual hostility. The generally defiant or indifferent outlook of new residents had decidedly negative results, as it permitted a high tolerance for the destruction of the architectural monuments left to them by the departed prewar population.«[52] The basis from which Tscherkes deduces the feelings of »postwar settlers« and the workings of their community is not clear since he does not refer to these people's testimonies. But it is also worth noting that this author does not single out the newcomers from the villages. Studying Stalinist urban planning he must have seen that the desire to change the old city was the desire of the state authorities; those who actually tried to destroy the old city were not villagers but specialists on cities, experts and, in a sense, arch-urbanites as concerned with the creation of a perfect city as many other modernist architects were.

The other side of the divide between urban and rural »culture« was also used for self-assertion. The so-called »villagers« told stories about uncultured barbarians who took over apartments and their contents without knowing what do with them. Stories about Soviet officers' wives going to the opera in nightgowns or washing their hands in toilet stalls were told by these new city dwellers about those who claimed not only social but also cultural superiority. They looked back to the vibrant and fairly modern culture in their pre-war villages that the Soviet authorities had destroyed: to the reading clubs, amateur performances, recitals and festivals, all of which were gone. In this version of the story the old city's decay was attributed not to the invasion of the »barbarians« from the villages in the 1970s – they went to the new areas and new buildings, anyway – but to the inability of the new »masters« to take care of them. The dying out of Polish gardeners, concierges and watchmen inevitably brought about the death of parks, front doors, walls and roofs.

I refer to these stories here not to refuel old animosities but to demonstrate that an entire rich layer of urban folklore has somehow been omitted in the publications on L'viv under discussion and that this, perhaps, indicates another purpose such an overemphasis of and speculation on the urban-rural divide serves. By opposing the rural and siding with the urban, Ukrainian intellectuals are trying to bypass old national divides; in this way they represent themselves as cosmopolitans whose motherland is not a country but a culture, who are flesh of the city's flesh and have nothing to do with the pastoral and rustic »Ukrainian national« tradition.

IV. European versus Soviet

While spatially the city under construction is defined against the village and the countryside, temporally it is defined against the Soviet period. Since the new urban myth is trying to depart from previous nationalist stereotypes, the Soviet period is not blamed for the suppression of nationality or its language policy. Rather it is blamed for the destruction of the elusive urbanity that had been there before the Soviets came. We have already caught a glimpse of this in Tscherkes's article, which mentions post-war residents' »high tolerance for the destruction of the architectural monuments.«[53] (In terms of concrete evidence it seems that the destruction of historical monuments has been going on at a much faster pace since Ukrainian independence than in Soviet times.)

Western Ukrainian intellectuals share a certain tendency in studies of the post-Soviet or post-Communist transition: an attempt to define the Soviet heritage so that it could be overcome. This tendency sees the solution to many post-Soviet problems in building a vibrant public sphere independent from the state. Some believe that the city is a crucial site for the formation of responsible citizens who are able to intervene. In Blair A. Ruble's collection on urban history and civic identities, it is argued that:

> The problems of a shared public civic identity that can embrace a variety of urban groups and individuals is especially troublesome in post-authoritarian states. Previous ideologies often have become so transparently discredited that they have long ceased providing meaningful markers for how people think about who they are. Perhaps even more problematic for building a democratic policy, Soviet-style socialist ideology in

> particular demanded an identity of interest between the totality of society and the Communist Party-dominated state.[54]

Ruble's co-editor John Czaplicka develops this argument further:

> the concept of ›civic‹ blends the political dimension of citizenship rights and responsibilities with the sense of belonging to a place and more specifically with the projection of a locality often associated with the city. This blend is devoid of exclusive and divisive ideological connotations that are inherent in concepts of identity associated with religion, ethnicity, race or nation.[55]

According to Czaplicka, Soviet urban space ignored the specificity of local, peopled streets with essentially similar monuments to anonymous workers, communist personages and soldiers. It was an attempt to eschew local history and to create a homogeneous space for the formation of *homo sovieticus*, but it did not succeed:

> The utopian paradigm of Soviet history proved unable to generate a new sense of belonging related to a sense of place so fundamental for constituting a local identity. In its form and idea the socialist city broke loose from any regional or local moorings, and its lack of specificity was not conducive to establishing identity with place. Furthermore, seen from the perspective of a resident of Riga, Vilnius, or L'viv, this official history in its full Soviet-Russian import was the history of an imperial state colonizing subject lands on its periphery.[56]

Czaplicka ends his overview of Soviet urban history in the region with the return in the past two decades of suppressed local history, a return grounded very much in architecture, urban planning and other similar remains of Europeanness that have survived on the western borderland of the Soviet Union.

Others perceive this belonging to Europe as natural, as proven by L'viv's architecture and the longings of its intellectuals: »the final years of Soviet rule, however, saw the return, after a six-decade absence, of a Central European community sneaking in under these artificial boundaries, reaching back across the formidable Soviet border to Ukrainian Galicia and, in particular, to Lviv.«[57] This particular article ends with a statement about »the path that has led Ukraine back to a place in Central Europe.«[58]

Several points need to be made here. First of all, the authors seem to subscribe to the view of Soviet society as totalitarian and tightly controlled by the state. This view has been challenged by many social historians of the Soviet period who have shown that people were not only passive victims of the regime, that their agency can be seen in many social and cultural processes that took place in the Soviet Union, that the regime used not only suppression but also negotiation, and that the state was secured not only by coercion but also by consent.[59] Ruble in another text mentions that the social experience of a Soviet city in many ways resembled that of any other: »Metropolitan life – at the height of capitalism, under Soviet-style socialism, or in various pre- and post-modern incarnations – by its very nature produces societies so complex and spatially dense that people of difference are forced by physical proximity to interact with neighbors quite unlike themselves.«[60]

Secondly, Soviet socialist ideology was hardly universally accepted, nor did it fully penetrate individual consciousness.[61] In fact from Soviet society can be induced the formation of a »people versus power-block« identity at its purest. Finally, as we have seen from previous discussion the distinction between the colonized and colonists in this context is also not that simple. At least some »colonists« claim to self-identify with the city and Europe in general more than the local »colonized« did. Finally, it should be mentioned that, in the context of contemporary Ukraine, »Soviet« has not only a temporal dimension but seems to be a very alive part of the contemporary cultural landscape. Large segments of the population, if not identifying with the »Soviet« heritage directly, nonetheless use it as part of their cultural arsenal and draw on it to articulate their grievances and concerns.[62]

All the connotations of Soviet mentioned above, and many more, show up in the statements of Western Ukrainian intellectuals. In her editorial to the *Ji* issue, Iryna Magdysh complains that the city is still threatened by this Soviet identity:

> A little sub-unit of a ›sovietized‹ world. This is also a pressing issue. The city has become fashionable. Donetskites, Dnipropetrovskites, …ites, …ites, in general Soviet [people] understandably tired from their gangster sorting-outs in native *luhansks*[63] unexpectedly discerned in the ›fearsome Banderite‹[64] Lviv, a quiet and beautiful European city, absolutely fit for aristocratic education of the descendants of a mixed breed of Krasnyi Luch [an industrial Donbass slum] and Huliaipole [Nestor Makhno's home town and the center of his activities during the Civil War, symbolic of the anarchist character of peasant Eastern Ukraine]. Also quiet in the sense that it does not resist the *surzhyk* [a vernacular mix of Ukrainian and Russian] so common among newcomers on its streets, the plastic windows in the secessionist buildings, the orgy of foreign-made cars with Kiev plates on the cobblestones. This is actually a list of what these newcomers have already brought to the city. Even worse is that such a style is easily adopted by young L'vivites not yet formed by the city environment. Taking as their example ›Kiev‹ or ›Dnepropetrovsk‹ [i.e. gangsters], they paint ugly graffiti on the buildings in Rynok square.[65]

Moreover, in this particular vision, L'viv is also threatened by the invasion of refugees and émigrés from »far Asian countries« – »just like some time in the past, the way to ephemeral Eden passes through this city, which is very real and more hospitable than Europe.«[66]

Obviously this vision of the besieged cultured city barely withstanding pressures from the Soviet east and the East in general is an extreme version of the new narrative. But motifs like those indicated above appear elsewhere. In the case of the newly constructed L'viv, real urban culture is European and not Soviet. In fact it is alien to the former Soviet culture. What remains of the post-Soviet mentality is quite often associated with pop- and mass-culture and also attributed to Russian culture.[67] Soviet-based identity is represented as semi-criminal. Counter-culture also takes on criminal and Soviet overtones. These representations have no basis whatsoever in terms of actual »empirical« analysis. We can take graffiti as an example. This is a Western phenomenon that came to Eastern Europe and L'viv in particular together with the fall of communism. L'viv graffiti has developed without any particular influence from Ukraine's central and eastern urban centers. The most popular graffiti style in L'viv, for example, »tribal« is hardly to be found in Eastern Ukraine. Poland, and not some country to the east, seems to be the real leader in European graffiti.[68]

The case of graffiti counter-culture, as well as pop-culture slummed together and labeled as Soviet remains that threaten real urban culture, indicates what kind of culture these myth-makers have in mind. Such a notion of »culture« was hardly foreign to the Soviet era and Soviet authorities, and was actually implemented in the special arrangements and architecture of Soviet cities. Kiev, for example, was rebuilt by the Soviet authorities according to their vision of a »well-ordered and cultured socialist metropolis,« in which architectural innovations were accompanied by »the reintroduction of ›bourgeois‹ habits of personal hygiene, leisure and entertainment,« state support for the institutions disseminating classical culture, opera in particular.[69] In its civilizing mission the Soviet state cultivated both everyday and high culture (from the basics of hygiene to classical opera) among its workers and peasants – »culture was something that had to be mastered, like virgin lands and foreign technology.«[70] How are the attitudes of the intellectuals propagating their version of urbanity as culture different from these? The only difference I see so far is that they believe that culture cannot be mastered either by villagers or by »Soviet« people – it has to be cultivated for generations. Is this supposed to be a step forward from the Soviet take on these matters?

V. The Culture and Society of Mythical Cosmopolitanism

In his discussion of contemporary Western academics constructing nations in Central-Eastern Europe and Asia Minor, Chris Han has noted that »rather than recognize culture as something that is continuously variable and blurred, with all attempts to impose boundaries that are ›fuzzy‹ at the edges, the intellectuals discussed here cling to the notion of the sharp boundary.«[71] On the surface this observation does not seem to apply to the mythic constructions discussed here. In his latest novel, L'viv's best-known contemporary writer explicitly locates the national with all its phantasmal but pleasurable narratives inside the city's dump.[72] The imagined urban culture transgresses national boundaries (or at least many of these, the problematic figure of the threatening »East« is still there). The boundaries that these new representations create are not the territorial boundaries of ethnic culture but the social boundaries created by the mechanisms of hierarchizing, subordinating and ordering implicit in these intellectuals' vision of culture. The workings of inclusion and exclusion here are based not on in-born belonging, language or religion but on the acquired ability to distinguish, to possess cultural/social capital in Bourdieu's sense. Exclusion in this case is not absolute othering associated with the drawing of a clear-cut boundary but social distancing and the assignment of lesser cultural value.

I have shown that the identity of the mythical city is based on multiple oppositions: between urban and rural, Soviet and European, civilized and barbarian. But there is also believed to be some core essence in this city culture; unique features that make it attractive are being defined. Ihor Chornovol describes his longings for his ideal city in this way: »I disliked boors, carols with wishes of sheep and calves in the apartment on the ninth floor. What was I missing and, unfortunately, am still missing today? Authenticity and the identity of the city.«[73] Chornovol believes that his identification with the urban »*Heimat*« is much better than identifying with the national »*Vaterland*« (he purposefully introduces these German words to back his views, missing the fact that *Heimat* is as overburdened with nationalist connotations as *Vaterland* is).[74] This locus of a new and allegedly better identity claims the supreme loyalty its nationalist predecessor used to enjoy and the vocabulary of this claim has not changed much – »L'viv *über alles*.«[75]

These longings are also articulated in the political programs of a politician who believes that »not countries, not transnational companies or supra-state unions will create future global world but cities.«[76] When it comes to the concrete image of the city he has in mind we get the following picture:

> We want our cities to be comfortable and cozy, just as L'viv was in the 1930s – a lot of banks, restaurants, hotels, unique urban folklore, good beer, tasty coffee, sweet chocolate, the whole spectrum of food-processing industries providing income for the city's

> coffers, luxuriant vegetation in parks, beautiful women, portly men, a lot of creativity and romanticism.[77]

This predominantly gastronomic utopia is very revealing as to the kind of culture and identity the myth-makers have in mind. This is the culture of the once lost and being recovered bourgeois world. Instead from imagining and carving up a new nation with better chances of getting into »Europe,« the concept of urbanity is taken and upheld as a more realistic and more realizable gateway to Europe. But the Europe in question is not contemporary European society; rather it is what Geoff Eley defines as European culture in the »more anthropological sense of aristocratic and bourgeois culture; most easily graspable through fashion, taste, and style – in food, dress, vacations, the social rhythm of the season, family relations, sexual mores, the international circuits of education, the accumulation of cultural capital, and so on.«[78] The irony is that Sadovyi locates his ideal in the 1930s when this bourgeois culture was less certain of itself than at any point before, when there was a widespread feeling of the city's decline and decay even in L'viv itself.[79]

Vozniak in his programmatic article on economic development offers a more future-oriented program, the typical program of an ideal post-industrial society, an Eden for intellectuals. According to this program, L'viv has to become a center of the »post-Soviet-US« borderland.[80] To do so the city has to get rid of »old industrial approaches« (as if the city had not got rid of them during the USSR's collapse). »The unquestionable priority should be the development of an educational structure«; L'viv has to become the hub of Ukrainian-Polish relations, and relationships between Ukraine and the European Union, and a center of Ukrainian language, culture, and of course, tourism.[81] Let's note that there is no mention in these programs of social problems and their possible solutions.

These bourgeois visions of the city and its culture are supported by the focus of most current research on L'viv. Czaplicka sums up this particular version of culture: »The emphasis in this collection is on culture, cultural production and processes, and the interrelationship of cultural distinctions with politics. Two essays on literature, three on architecture and urban planning, and one on the cultural competition between the Poles and the Ukrainians of Lviv are meant to provide the reader with an understanding of the city as a site of unique cultural production.«[82] This summation reveals the emphasis on both high culture and on the national narratives struggling in and for the city. There is very little of the declared need to »look closely at local history.«[83] While in the case of other Central European cities there is a tendency in urban history to go beyond the stereotypical and very narrow definition of these cities' respective »cultures,« nothing like this has taken place in the case of L'viv.[84]

Czaplicka claims that »The ›multicultural‹ character of the city [...] is a leitmotif throughout this collection [...]«[85] But the multiculturalism in question is limited to the multiplicity of ethnicities or, to be precise, of national projects and thus seems to reinforce national stereotypes that reduce cultural difference to ethnic or national. While the architecture in question is defined »as a unique document that embodies the spirit of past ages and reflects the varying aspects of the pluralistic culture of the western region of Ukraine on the border with Central Europe,«[86] the collection does not mention those inhabiting the space, maneuvering in and appropriating it. Even Hrytsak, trying to counter the narratives that condemn the post-World War II Ukrainian population of L'viv, points to the existence of »high« Ukrainian culture preserved by the vestiges of the pre-war Ukrainian intelligentsia, thus accepting these narratives' disregard for the »low« and plebeian.[87] Is it a slip of the pen that Scholem Alejchem (who was born in the Russian Empire, died in New York and in most of his works, written largely in Yiddish with some in Russian, described Jewish life in the Pale of Settlement) is referred to as a »Jewish-Galician« writer?[88] Or does it betray a particular imaginary in which multiculturality ends at the western border of the Habsburg Empire? In no sense was the Russian Empire just across the border less »multicultural« than Galicia itself, or for that matter the Soviet Union.

The question also arises of why these implications of the new urban imagination developing in L'viv have not been noticed by Western intellectuals. As Markus Reisenleitner has shown, many of those working with the larger post-Habsburg space tend to view it as a ready ground for the application of a fashionable theoretical toolbox without much reflection on the implications of such applications.[89] In a way similar to the local intellectuals, Western intellectuals seem to prefer to engage in the construction of imagined spaces of belonging and location and not their deconstruction. On the one hand, some seem to believe in the mechanistic self-sorting associated with capitalism and think that the return of capitalism will eventually lead to liberalization, negotiation, hybridity and tolerance. Blair A. Ruble, for example, expresses his faith in capitalism's liberating powers in the following way: »Capitalism, while concentrating power, wealth and resources in a few hands, also undermines the capacity of any single group to establish durable dominance over others. This is especially so at those moments when capitalist modes of production come into being.«[90]

Others simply see cities as sociologically destined to become a cosmopolitan alternative to the nation-state. Boym believes that »In the current opposition between global and local culture, the city offers another alternative – that of local cosmopolitanism. This kind of cosmopolitanism is not based on electronic interface but on face-to-face cross-cultural encounters of strangers in a physical space.« She notices that »In some cases, such as in Prague or St. Petersburg, urban cosmopolitanism is not a feature of the present but rather an element of nostalgia« but nevertheless believes that »yet it serves to underscore both the global and national discourse.«[91] It is forgotten that certain cosmopolitan visions can produce high modernist visions as destructive to local practices as any national project.[92] The scale of the project does not automatically translate into local either.

Some disturbing implications of similar projects have been brilliantly revealed by Pamela Ballinger in her work on Istria, Triest in particular, and the place's claims to a hybrid identity.[93] At first eagerly subscribing to the rhetoric of the new identity project which claims Istria as a particular place that has been hybrid for ages, with time Ballinger recognized both the nineteenth-century bourgeois lineages of the project and the fact that in this case the discourse of hybridity actually reinstated the map of a world inhabited by homogeneous bounded cultures, the interaction between which cross-fertilizes only some of the places in question. All of this holds true for L'viv as well, except that intellectuals do not use the term hybridity as they are satisfied with the term multiculturalism and its implications of different cultural strands coexisting in the same space.

In the nineteenth-century bourgeois world, the urban identity of a place as part of civic culture was ripe with the connotations of civilization and served as an instrument of social control.[94] Multiple exclusions were not limited to popular or low-brow, lower class, »marginal people from bedroom communities.« As in traditional bourgeois narratives, this version of culture also excludes on the basis of age and gender.[95] It also seems that some of the exclusions mentioned here can become fundamental, constitutive of the new *civitas* being imagined, in the same way some fundamental exclusions became constitutive of the bourgeois public sphere in the nineteenth century.[96] Instead of appreciation for the fragmented and fluctuating geography and identities of the city, this intellectual project betrays its longing for the city unified and ordered. Instead of decentered urban ethnography, its knowledge claims are based on the construction of yet another canon serving a functional understanding of a centered and coherent society.

At the center of this perfect society the intellectuals locate civil society (they are its foremost members, of course), which they fashion according to their taste and which in many cases coincides formally with the organizations and networks they have created. Quite often this civil society is used to underline differences and to prove the existence of different patterns of societal interaction and different kinds of society in the case of, for example, L'viv versus eastern Ukraine.[97] A parallel with the postcolonial context may be drawn here. Partha Chatterjee has warned us that it might be too early to discard the nation-state and to claim moving »beyond« it. In the postcolonial context, civil society, which is invoked as a global alternative to the nation-state, very exclusively embodies the desire of local elites »to replicate in its own society the forms as well as the substance of Western modernity.«[98] On the other hand, a political society whose boundary coincides with the boundaries of the nation-state remains more open and seems to constitute a space in which divergent tendencies may co-exist and struggles for democracy can be fought. Symptomatically, the political mobilization and street protests in Ukraine at the end of 2004 occurred because of concerns about democracy and violations of democratic procedures by the state institutions while democracy has never figured among the major concerns of the Ukrainian intellectuals under discussion, who are more concerned with liberal values, Culture, civil society, its institutions, social capital and so on.

To conclude, I would like to return to the connection between academic writing and values. Yaroslav Hrytsak has written that »writing a multicultural history becomes more than an academic exercise; to some extent, its success or failure might have an impact on the political future of both L'viv and Ukraine, and, in a larger context, of the whole Central and Eastern European region.«[99] I agree with this statement wholeheartedly since I believe that the writing of history has always been »more than an academic exercise.« But the notion of »multicultural history« must be expanded to pay attention to class, gender, youth, those with disabilities and many other lines along which difference is produced. Attention should be paid not only to the normative ideal intellectuals have in mind but also to the actual tensions, conflicts, struggles and negotiations that occur on the ground.

NOTES

1 For an example of two nation-construction projects enlivened by intellectuals who, unlike their nineteenth-century counterparts, are quite conscious that their activity is not reviving but constructing nations and who use for their projects the toolbox identified by scholarship on national identity and movement, see Chris Hann, »Intellectuals, Ethnic Groups and Nations: Two Late-Twentieth-Century Cases,« *Notions of Nationalism*, ed. Sukumar Periwal (Budapest-London-New York: Central European University Press, 1995), 106-28.

2 Examples can be found in Ronald Grigor Suny and Michael D. Kennedy (eds.), *Intellectuals and the Articulation of the Nation* (Ann Arbor: University of Michigan Press, 1999).

3 Andriy Zayarnyuk, »At the Frontiers of Central Europe: Ukrainian Galicia at the Turn of the Millennium,« *SpacesofIdentity* 1, 1 (2000): 15-34, <http:// http://www.univie.ac.at/spacesofidentity/Vol_1/_PDF/Zayarnyuk.pdf> accessed on 25 February 2005.

4 Andrii Sadovyi, »L'viv – i my u L'vovi,« *Postup*, 24, 21 February 2002.

5 OCLC World Catalogue's search gives the following statistics for books in Ukrainian with »Lviv« as subject phrase: 1989-1994: 32 books, 1995-1999: 47, and 2000-2004: 67.

6 *Strasti po L'vovu. Krytyka*, 6, 7-8 (2002): 57-8.

7 John Czaplicka (ed.), *Lviv. A City in the Crosscurrents of Culture*, special issue of *Harvard Ukrainian Studies* 24 (2000, printed in 2002).

8 »L'viv, Leopolis, Lwow, Lemberg: Genius Loci.« *Ji* 29 (August 2003, printed in 2004) (further on – simply *Ji*). The goal of the NGO is described on <http://www.ji-magazine.lviv.ua/vidhuky/about-ukr.htm> accessed on 25 February 2005.

9 *Austrian History Yearbook* XXXIV (2003): 57-110.

10 Further on – simply *Krytyka*.

11 This was acknowledged by *Krytyka*'s editor Andrii Mokrousov during the issue launch held in L'viv in 2002.

12 The term became extremely popular in the representations created by these intellectuals. Figuring in the very title of the above-mentioned special issue of *Ji*, it is used repeatedly by the authors represented there. Cf. also Ihor Chornovol, »Genius Loci,« *L'vivs'ka Hazeta* 2 (328), 9 January 2003. In all these cases the phrase describes the substance of a specific and, at the same time, universally urban socio-cultural climate that L'viv has lost.

13 This has been sketched out by Yaroslav Hrytsak, »Strasti po L'vovu,« *Krytyka*, 6.

14 Yaroslav Hrytsak's speech at the launch of George G. Grabowicz, *Do istoriï ukraïns'koï literatury: doslidzhennia, eseï, polemika* (Kyïv: Krytyka, 2003), Ukrainian Catholic University, Lviv, 2003.

15 George W. Grabowicz, *Shevchenko iak mifotvorets': semantyka symvoliv u tvorchosti poeta* (Kyiv: Radians'kyi pys'mennyk, 1991). Original: George G. Grabowicz, *The Poet as Mythmaker: A Study of Symbolic Meaning in Taras Shevcenko* (Cambridge: Harvard University Press, 1982).

16 Even the nationalist Union of Ukrainian Youth now organizes workshops on national myth-making. Information from Yaroslav Hrytsak's speech at the launch of George G. Grabowicz, *Do istoriï ukraïns'koï literatury: doslidzhennia, eseï, polemika* (Kyïv: Krytyka, 2003), Ukrainian Catholic University, Lviv, 2003.

17 George G. Grabowicz, »Mythologizing Lviv/Lwów: Echoes of Presence and Absence,« *Lviv. A City in the Crosscurrents of Culture*, special issue of *Harvard Ukrainian Studies* 24

(2000): 313-42. Unfortunately his article concentrates on the national narratives about L'viv of longer duration and touches upon current myths only towards the end.

18 Ibid., 312.

19 Bronislaw Malinowski, »In Tewara and Sanaroa – Mythology of the Kula,« in *Malinowski and the Work of Myth*, selected and introduced by Ivan Strenski (Princeton: Princeton University Press, 1992), 13.

20 Ihor Mel'nyk, »Ostanni l'vivs'ki dekady XX-ho stolitta, abo Tlo poetychnoho mifu,« in Nadiia Morykvas, *Za nas – u L'vovi: Mity i mify* (L'viv: Piramida, 2001), 148. The book itself is the best testimony of the »myth's« currency in the variegated circles of L'viv intellectuals. Grabowicz discards this book as a product of commercial opportunities connected with L'viv's popularity, »a pretentious salad of personal reminiscences and anecdotes raised now to the level of ›legend‹ and ›myth‹ […] not deserving particular attention – except for the way in which it reminds us that no topic is proof against banality, and that the essence of the Lviv theme, as of any other cultural narrative inheres in the tone and the perspective, not in the manifest content.« Grabowicz, »Mythologizing Lviv/Lwów: Echoes of Presence and Absence,« 337. What underpins his statement is his view of myth as a necessarily elaborated, rhetorically powerful and aesthetically attractive structure. However, Grabowicz misses the fact that this book served the authors of a particular generation and milieu as a statement of their participation in the intellectual life of the city. Indeed they played quite a prominent role in the public life of the city at the end of the 1980s-early 1990s and Ihor Mel'nyk himself served as a source of knowledge about these events for another author in this issue of *Harvard Ukrainian Studies* – see Padraic Kenney, »L'viv's Central European Renaissance, 1987-1990,« *Lviv. A City in the Crosscurrents of Culture*, special issue of *Harvard Ukrainian Studies* 24 (2000), 312, fns. 11 and 12.

21 Pierre Hegy, *Myth as Foundation for Society and Values. A Sociological Analysis* (Melen Studies in Sociology, v.10) (Lewiston, Queenston/Lampter: The Edwin Mellen Press, 1991).

22 <http://www.ji-magazine.lviv.ua/seminary/2002/sem27-06.htm> accessed on 25 February 2005.

23 Iurii Prokhas'ko, »Kakaniia chy Tsekaniia? (pro ›misiiu Halychyny‹),« *Krytyka.*

24 Aleksandra Matyukhina, *W Sowieckeim Lwowie. Życie codzienne miasta w latach 1944-1990* (Kraków: Wydawnictwo Uniwersytetu Jagiellońskiego, 2000), 149-68 – chapter on

»Mit Polski w sowieckiem Lwowie.« For a different take on a similar problem and, I believe, heuristically more useful terminology, see William J. Risch, »Thinking between Borders: Polish Media and Cultural Resistance in post-1953 L'viv,« forthcoming in *Approaching Ukrainian Identities*, special issue of *Canadian American Slavic Studies*, 40, 1 (Spring 2006).

25 Iaroslav Hrytsak and Viktor Susak argue that after the Second World War »The salient issue was to integrate a rural population that had no strong national myths of urban experience.« Yaroslav Hrytsak and Victor Susak, »Constructing a National City: The Case of L'viv,« in *Composing Urban History and the Constitution of Civic Identities*, eds. John J. Czaplicka and Blair A. Ruble (Washington, DC: Woodrow Wilson Center Press, 2003), 140-1. Again it is not clear for whom this was an issue back then, unless some transcendental city existed through the years with an agenda of its own. For the Soviet authorities the issue was to integrate the rural population into Soviet society and culture and they did it in the 1920s and 1930s without resorting to the »myths of urban experience.«

26 See, for example, »The True History of the Kingdom of Galicia« published in *Postup* throughout 2003 – Stas Roland-Perfets'kyi, »Spravzhnai istoriia korolivstva Halychyny,« the first installment appeared in *Postup*, 52, 5 April 2003.

27 I have in mind in particular Vasyl' Kozhelianko, *Defiliada v Moskvi* (L'viv: Kal'variia, 2000). This »Parade in Moscow« envisions a future in which the Ukrainian army in alliance with Nazi Germany defeated the Soviet Union, then defeated totalitarian Nazi Germany and finally established a new democratic order under Ukrainian aegis. This alternative future is juxtaposed with Ukrainian realities full of the colonial heritage of subordination to everything Russian.

28 Blair A. Ruble, »Living Apart Together: The City, Contested Identity and Democratic Transition,« in *Composing Urban History and the Constitution of Civic Identities*, eds. John J. Czaplicka and Blair A. Ruble (Washington, DC: Woodrow Wilson Center Press, 2003), 1.

29 Ibid., 14.

30 Ibid., 20.

31 The attempt to add a crown to the city's coat of arms and the heated discussion around this issue connected with old national sensibilities is an example of an intervention aimed at changing public representations of the city. <http://www.ji-magazine.lviv.ua/seminary/2003/sem03-07.htm> accessed on 25 February 2005.

32 Katherine Verdery, »Civil Society or Nation? ›Europe‹ in the Symbolism of Romania's Postsocialist Politics,« in *Intellectuals and the Articulation of the Nation*, eds. Ronald Grigor Suny and Michael D. Kennedy (Ann Arbor: University of Michigan Press, 1999), 393.

33 Ibid., 394.

34 Daniel Kelliher, »Keeping Democracy Safe from the Masses: Intellectuals and Elitism in the Chinese Protest Movement,« *Comparative Politics* 25, 4 (July 1993): 379-96.

35 Svetlana Boym, *The Future of Nostalgia* (New York: Basic Books, 2001), *passim*.

36 Taras Vozniak, »L'viv. Sine qua non – bez choho nemaie,« *Ji*, 395-8.

37 Ibid., 401, italics in original.

38 Aleksandra Matyukhina, *W Sowieckeim Lwowie. Życie codzienne miasta w latach 1944-1990* (Kraków: Wydawnictwo Uniwersytetu Jagiellońskiego, 2000), 38.

39 Andrew Lees, *Cities Perceived. Urban Society in European and American Thought, 1820-1940* (New York: Columbia University Press, 1985), 1-2.

40 Peasantry under capitalism is not the same class as peasantry under feudalism. For an introduction to this discussion see Hamza Alavi, »Peasantry and Capitalism: A Marxist Discourse,« in *Peasants and Peasant Societies: Selected Readings*, ed. Teodor Shanin (Oxford: Blackwell, 1987), 185-96.

41 Vozniak, »L'viv. Sine qua non – bez choho nemaie,« 401.

42 Ibid., 397.

43 Ihor Chornovol, »Bastion trymaiet'sia,« *Krytyka* 1–2 (2004): 75–76.

44 Uffe Østergård, »The Danish Path to Modernity,« *Thesis Eleven* 77 (May 2004): 25-43. Some have argued that the strength and cohesion of village communes in Western Europe helped the peasants there to repel the assault of feudal seigniors preventing the rise of the »second edition of serfdom,« which transformed Eastern Europe into the periphery of modern capitalism. Robert Brenner, »Agrarian Class Structure and Economic Development in Pre-Industrial Europe,« *Past and Present* 70 (Feb. 1976), 30-75.

45 Vozniak.

46 Grigorii Komskii, »Teni tenei,« *Ji*, 360.

47 Ibid.

48 Il'ko Lemko, *L'viv ponad use* (L'viv, 2003), 139.

49 Maciej Janowski, »Lemberg 1916 – L'viv 2000. Rozdumy nad starym putivnykom,« *Krytyka*.

50 Komskii, 360.

51 The nationalist hatred in Igor' Klekh's representations is analyzed in Yaroslav Hrytsak, »Strasti po L'vovu,« 4.

52 Bohdan Tscherkes, »Stalinist Visions for the Urban Transformation of Lviv, 1939-1955,« *Lviv. A City in the Crosscurrents of Culture*, special issue of *Harvard Ukrainian Studies* 24 (2000), 206.

53 Ibid.

54 Blair A. Ruble, »Living Apart Together: The City, Contested Identity and Democratic Transitions,« in *Composing Urban History and the Constitution of Civic Identities*, eds. John J. Czaplicka and Blair A. Ruble (Washington, DC: Woodrow Wilson Center Press, 2003), 2.

55 John J. Czaplicka, »Conclusion. Urban History after A Return to Local Self-Determination – Local History and Civic Identity,« in *Composing Urban History and the Constitution of Civic Identities*, eds. John J. Czaplicka and Blair A. Ruble (Washington, DC: Woodrow Wilson Center Press, 2003), 375.

56 Ibid., 383.

57 Padraic Kenney, »L'viv's Central European Renaissance, 1987-1990,« *Lviv. A City in the Crosscurrents of Culture*, 303.

58 Ibid., 311.

59 For a short overview, see Mark von Hagen, »Empires, Borderlands, and Diasporas: Eurasia as Anti-Paradigm for the Post-Soviet Era,« *American Historical Review* 109, 2 (Apr. 2004): 445-69.

60 Blair A. Ruble, *Second Metropolis: Pragmatic Pluralism in Gilded Age Chicago, Silver Age Moscow, and Meiji Osaka*. Washington: Woodrow Wilson Center Press, 2001, 29.

61 For a sophisticated treatment of this problem, see Alexei Yurchak, *The Cynical Reason of Late Socialism: Language, Ideology and Culture of the Last Soviet Generation* (Ph.D. dissertation, Duke University, 1997).

62 Warner, elections.

63 This common noun is created from the proper noun »Luhans'k« – administrative center in the Donbass region.

64 The adjective »Banderite« was created from the surname of Stepan Bandera (1909-1959), the leader of the Organization of Ukrainian Nationalists.

65 Iryna Magdysh, »L'viv. Genius Loci,« *Ji*.

66 Ibid.

67 The recent and widely responded to example of this was »Open Letter of Twelve Apolitical Writers on Choice and Elections,« *Smoloskyp*, #10(111), October 2004, <http://www.smoloskyp.kiev.ua/docs/bul_10_04.htm#41> accessed on 25 February 2005.

68 Based on the interview with graffiti writer Iurii (tag AK47) by Nadiya Chushak, courtesy of Nadiya Chushak.

69 Serhy Yekelchyk, »The Making of a ›Proletarian Capital‹: Patterns of Stalinist Social Policy in Kiev in the mid-1930s,« *Europe-Asia Studies* 50, 7 (1998): 39-40.

70 Sheila Fitzpatrick, *Everyday Stalinism. Ordinary Life in Extraordinary Times: Soviet Russia in the 1930s* (New York: Oxford University Press, 1999), 79.

71 Hann, »Intellectuals, Ethnic Groups and Nations,« 123.

72 Iurii Vynnychuk, *Mal'va Landa* (L'viv: Piramida, 2003).

73 Chornovol, »Genius Loci.«

74 Ibid.

75 This is the translation of the Ukrainain title of Lemko's book *L'viv ponad use.*

76 Andrii Sadovyi, »Poklykannia mista v epokhu globalizatsiï,« *Postup*, #114, 3 July 2003.

77 Ibid.

78 Geoff Eley, »Culture, Britain, and Europe,« *The Journal of British Studies* 31, 4 (October 1992): 394.

79 Anna Veronika Wendland, »Post-Austrian Lemberg: War, Commemoration, Interethnic Relations, and Urban Identity in L'viv, 1918-1939,« *Austrian History Yearbook* 34 (2003): 83-102. Also Andrzej Bonusiak, *Lwów w latach 1918-1939: Ludność, Przestrzeń, Samorząd*, (Galicja i jej dzidzictwo, t. 13) (Rzeszów: Wydawnictwo WSP, 2000).

80 Vozniak, »Sine qua non,« 400.

81 Ibid., 404.

82 Czaplicka, »Lviv, Lemberg, Leopolis, Lwów, Lvov: A City in the Crosscurrents of European Culture,« 18.

83 Ibid., 17.

84 For example, Wolfgang Maderthaner and Lutz Musner, *Die Anarchie der Vorstadt: das andere Wien um 1900* (Frankfurt/Main; New York: Campus, 1999), or Cathleen M. Giustino, *Tearing Down Prague's Jewish Town: Ghetto Clearance and the Legacy of Middle-Class Ethnic Politics around 1900* (New York: Boulder, 2003).

85 Czaplicka, »Lviv, Lemberg, Leopolis, Lwów, Lvov: A City in the Crosscurrents of European Culture,« 23.

86 Tscherkes, »Stalinist Visions,« 206.

87 Iaroslav Hrytsak, »Strasti po L'vovu.«

88 Czaplicka, »Lviv, Lemberg, Leopolis, Lwów, Lvov: A City in the Crosscurrents of European Culture,« 37.

89 Markus Reisenleitner, »Central European Culture in Search of a Theory, or the Lure of Post/colonial Studies,« *SpacesofIdentity* 2, 2 (2002) <http://www.univie.ac.at/spacesofidentity/_Vol_2_2/_HTML/Reisenleitner.html>, Markus Reisenleitner, »Slashing Postcolonial Studies, or: Why this Debate still Bothers Me: A Response to Clemens Ruthner's »K.u.K. ›Kolonialismus‹ als Befund, Befindlichkeit und Metapher,« *SpacesofIdentity* 3, 1-2 (2003) <http://www.univie.ac.at/spacesofidentity/_Vol_3_2/_HTML/Reisenleitner.html>

90 Blair A. Ruble, *Second Metropolis: Pragmatic Pluralism in Gilded Age Chicago, Silver Age Moscow, and Meiji Osaka* (Washington: Woodrow Wilson Center Press, 2001), 18-19.

91 Boym, *The Future of Nostalgia*, 76.

92 James C. Scott, *Seeing Like a State: How Certain Schemes to Improve the Human Condition Have Failed* (New Heaven: Yale University Press, 1998), *passim*.

93 Pamela Ballinger, »›Authentic Hybrids‹ in the Balkan Borderlands,« *Current Anthropology* 45, 1 (Feb. 2004): 31-49.

94 Jon Stobart, »Building an Urban Identity. Cultural Space and Civic Boosterism in a ›New‹ Industrial Town: Burslem, 1761-1911,« *Social History* 29, 4 (November 2004): 485-98.

95 Not accidentally voices that challenge now dominant representations of the city by myth-makers and appear in less privileged spaces of writing (like walls and internet sites) are very often both youthful and female. See, for example, the apology of *Sykhiv* and a very different picture of the city space by livejournal's user »Zumka« – »Moï prostory,« <http://www.livejournal.com/users/zumka/> accessed on 2 February 2005.

96 Nancy Fraser, »Rethinking the Public Sphere: A Contribution to the Critique of Actually Existing Democracy,« in *Habermas and the Public Sphere*, ed. Craig Calhoun (Cambridge, MA: The MIT Press, 1992), 109-42.

97 Such an application of »social capital« based on the work of Robert Putnam was criticized in Dario Gaggio, »Do Social Historians Need Social Capital?« *Social History* 29, 4 (Nov. 2004): 499-513.

98 Partha Chatterjee, »Beyond the Nation? Or Within,« *Social Text* 16, 3 (Fall 1998): 62.

99 Yaroslav Hrytsak, »Lviv: A Multicultural History through the Centuries,« *Lviv. A City in the Crosscurrents of Culture*, 62.

Probe III:

The Centrality of Europe in the Global Circulation of Culture

Foreign Sound as Compensation: Social and Cultural Factors in the Reception of Western Music in Meiji Japan (1867-1912)

TAKENAKA TORU

I.

It was in the middle of the nineteenth century that the Japanese became acquainted with Western music. As the country broke with its long policy of seclusion and opened up to the outside world, many things rushed in from the West, and music was among them. In the westernization that set in, some were rapidly accepted, others not. It is noticeable that the Japanese felt almost unanimously estranged by Western music in those days. We hardly find a positive remark or reaction among those who had a chance to come into contact with it, for example, a senior member of the mission sent by the Shogunate government to Washington in 1860. He almost burst into laughter at a young woman's singing to a piano accompaniment when an American official entertained the mission with music at his home in Hawaii. The Samurai compared the soprano voice to that of a dog howling at night.[1] For ordinary people as well, Western sound was anything but pleasant. A British touring operetta, which dared a performance together with kabuki actors in a theater in Tokyo in 1879, met with a storm of grumbling and bawling from the audience. The artists did not sound to it like humans but rather like »hens being strangled.«[2] Even later, when music was compulsorily introduced into the elementary school curriculum at the turn of the century, pupils toiled at foreign melodies in the classroom. They »blushed for straining« to follow their teacher at the organ, as a contemporary observed.[3]

This Japanese lack of understanding is not very surprising. They had long been cut off from Western musical culture. The church music the Catholic missionaries had brought with them in the sixteenth century had soon fallen into oblivion due to the ban on Christianity. The subsequent national isolation had kept them free from outside influence for centuries. In addition, their native musical culture was unique. So the Japanese originally had no idea of the chord. Their sense of rhythm was based on the length of the notes and differed widely from the beating one in the West. The scale of Japanese traditional music consists of only five notes, whose height can be only approximately determined by the twelve notes of the Western temperament. In short, the native sound was so distinctive that it cannot be, to speak with a contemporary composer, transcribed on staff notation or played on the piano at all.[4] Furthermore, the Japanese always saw music as an accompaniment to or extension of the text, whether in the theater or with a storyteller, rather than as an independent object of aesthetic appreciation. That is why the genre of *katari-mono*, where the players recite rather than sing, occupies a good part of Japanese traditional music and why instrumental music by itself hardly developed in Japan.[5] It is, therefore, no wonder that the songs from the West should have sounded as strange as bestial cries to Japanese ears.

How large the difference was can be confirmed from the other side, too: the Westerners. Hardly any foreign visitors grew to like or appreciate the Japanese sound. The German doctor Erwin Baelz, who stayed in Japan from 1876 for almost thirty years and was the personal doctor to the Tenno, had great difficulty becoming accustomed to it. He shuddered at the »squeaking voice« of geishas and had to ask himself, listening to the *shamisen*, whether this was music at all. The aristocrat Ottmar von Mohl, who was invited from Prussia in 1887 to westernize the ceremonies at the imperial court, complained of »the screaming dissonance« of the traditional court music. Georg Michaelis, a teacher at a German school in Tokyo from 1885 to 1889 and the Reich Chancellor during the First World War, characterized the singing of the natives as squashing the voice »as if the singer would feel sick.«[6] Their remarks weigh all the more as they were by no means Japanphobes; they had otherwise high esteem for the native arts and crafts and praised temples and shrines in Kyoto and Nara without reservation.

We can thus confirm the unquestionable difference between Japan and the West in terms of musical culture. It is astonishing, therefore, how quickly the Japanese became accustomed to the strange sound from the West in the subsequent decades. And they went beyond simply growing accustomed to it; they even adopted it eagerly as if they had liked it from the very beginning. We are tempted to ask: How was the initial aversion resolved? Why did the Meiji Japanese make such efforts to accept the imported music?

While fully involved in discovering detailed facts about the development of Western music in Japan, music historians have hardly paid attention to these questions.[7] It seems that they implicitly assume Western musical culture to be humanity's supreme achievement. In their view it is the inevitable course of history that various native cultures eventually converge on it. They therefore find it no wonder that a wish for the westernization in music should have arisen in natives once they came into contact with European culture. From their standpoint Meiji Japan was a remarkably successful case because Japan made enormous progress in adopting it within a few decades. The aversion by the Japanese at the start makes the success all the more grandiose as the difficulties to be overcome were so grave. This view, biased by Eurocentrism and a mono-linear historical understanding, hardly satisfies us. It is not surprising that our questions lie beyond the scope of most music historians. Recently some historians have addressed our questions in the context of nation-building in the Meiji era.[8] According to their view, the government aimed to standardize musical life by means of the imported Western music in order to forge a culturally homogenized nation under the centralized power. So Western music was forced upon the people from above and became widespread. Admittedly, the role of government policy in introducing Western music was decisive. If the authorities hadn't introduced music into schools, the vast majority of Japanese would never have had any contact with it. But laws and decrees alone cannot change popular sentiment. Without spontaneous willingness on the part of the population, Western music would surely have failed to take root.

This article aims to examine the motives that led the Meiji Japanese to accept the foreign music. I do not focus on the individual psychology of the people involved but rather will make clear the social and cultural historical background.

II.

Roughly speaking, Western music entered Japan through four channels. First, Christian missionaries brought church music when they came to foreign settlements like Yokohama in the middle of the 1850s. With Protestant churches from the United States as a spearhead, the missionaries gained a further foothold after the ban on Christianity was lifted nationally in 1873. Japanese believers sang hymns together, and some learned to play the organ as well. The second channel was military bands. At the end of the Shogunate era, many *han* westernized their military system and introduced parading as part of it. For this purpose, for example, the Satsuma *han* dispatched a group of navy soldiers to Yokohama in 1869 to have them trained there. All of the *han*'s bands were later taken over by the imperial military. Beside their military service, the bands occasionally performed in public.[9] The third and the fourth channels were closely intertwined in the course of history. Court musicians were the third conveyors of Western music. In parallel with their hereditary occupation of providing ceremonial music, they were ordered to master Western instruments by the court authorities. The government now found it necessary to present the Tenno as a European-type sovereign, particularly on diplomatic occasions. Musicians were sent to the Institute of Music (*Ongaku Torishirabe Kakari*), the fourth channel, for training. The institute, established by the Ministry of Education in 1879 to introduce music into the public education system, examined pedagogical methods, produced teaching materials and trained schoolteachers. It was upgraded in 1887 to the Music Academy of Tokyo (*Tokyo Ongaku Gakko*) to be again reorganized as the Tokyo National University of Fine Arts and Music after the Second World War. It is thus the fourth channel that grew to be the pillar of Western music in modern Japan.[10]

To investigate the Meiji Japanese's motives for adopting Western music, the profile of those who not only listened to the newly introduced music passively but also learned to make this music actively will give us clues. Of these four channels, some information is available about the people who acted in the fourth. Let us begin with a list of the 22 students who were admitted to the Institute of Music in 1880 when it advertised its courses publicly for the first time.[11] With the exception of eight court musicians who were ordered to take part in the course, one was an aristocrat and eight were former Samurai (*shizoku*), while the other five came from the commoner (*heimin*) class. As far as their birthplace is concerned, ten came from the Tokyo prefecture and the other four from its neighboring prefectures. We have some information as well about the students of the Music Academy. Of the 17 students who graduated in 1897, seven were *shizoku* and the rest *heimin*, with the majority coming from Tokyo (five) and one or two from some of the other prefectures.[12] There are also records of the birthplaces of the twenty graduates of the previous year. The top is once again Tokyo; six came from the capital. The others are distributed more or less evenly.[13] The list of the students admitted in 1893 tells almost the same story. Five of 24 came from Tokyo, followed by three from Pref. Tottori and the rest from various prefectures.[14] We have to admit that the data available are by no means adequate, and the fact that relatively many came from Tokyo might not be very significant when we consider that the capital had attracted newcomers from the provinces and that all of the students whose registered birthplace was Tokyo were not necessarily born and bred Tokyoites. Nevertheless, we can still point out two outstanding facts. First, the former Samurai were clearly over-represented among the Western musicians in those days. Their share is around half of all the students in each of the above lists, which is strikingly higher than the national average (ca. 5%). Second, the geographical distribution of the students' birthplace is characteristic. Most of the prefectures in the lists were former *han* that either had been directly subordinate to the Shogun in feudal times (*shinpan* and *fudai*) or had taken his side in the civil war in 1868-1869 against the Tenno's government. On the contrary, there were very few who came from the so-called *hanbatsu* prefectures that mainly made up the leadership of the Meiji regime.[15] In the list of 1897 we find only two from Pref. Kochi, in that of 1896 none.

This is indicatively also proven by the distribution of the readership of a music journal. The mail subscribers to the first journal devoted to music in Japan, *Music Journal* (*Ongaku Zasshi*), were relatively concentrated in eastern prefectures like Miyagi and Nagano, apart from such prefectures with a large city in it like Tokyo, Osaka, Kyoto, and Aichi.[16] It was in eastern Japan, roughly speaking, that the pro-Shogunate powers had their stronghold and where major battles were fought in the civil war. By the end of the Meiji era, the geographical pattern had not changed. In 1909, of the 52 graduates of the Music Academy 40 came from eastern regions like Kanto, Koshinetsu and Tohoku. Correspondingly, those from the west made up a minority; as for the *hanbatsu* prefectures we find only two from Pref. Yamaguchi, and 20 had a Samurai family background, an extraordinarily high share.[17]

To sum up, we may assume that the Western musicians in the Meiji era were mainly recruited from the *shizoku* class in formerly pro-Shogunate regions, the »losers of the Restoration« so to speak, who suffered greatly from the collapse of the ancient regime. It is interesting that music did not stand alone in this respect at that time; in other genres of art, too, we meet with similar facts. While many craftsmen who relied on traditional skills in their work were artisans' offspring, there were many *shizoku* among the artists of Western painting; particularly in its pioneer generation many ex-Samurai of the Shogun's or of the *han* subordinate to him can be found.[18] The same holds true for literature. There were hardly any novelists or poets who belonged to the Satsuma or the Choshu clan, the two leading *hanbatsu* groups. Rather, the modern literature of Meiji, which broke drastically with the tradition that had reigned since the Edo era and tried to make a new epoch under strong Western influence by e.g. using colloquial style in the novel and adopting the Naturalist idea of description, was, as a critic says, completely produced by the people who had fought for the Shogun.[19]

Speaking of the appeal to the non-*hanbatsu* warriors, we remember another example outside art: Christianity. It is well known that the followers the missionaries converted first in Japan were mostly *shizoku*, above all the ones from the formerly pro-Shogunate camp. It is noticeable that many Christians can be found in the first generation of Western musicians, such as Nosho Benjiro, Okano Teiichi, Uchida Kumetaro, Nagai Koji, etc. Taki Rentaro, the first Japanese composer to work in the Western style to effect, converted shortly before he went to Leipzig to study in 1901. Many, too, had family or school surroundings in their youth in which Christianity played a role, even if they themselves were not of the faith. Yamada Kosaku, the most renowned Japanese composer and the founder of the first orchestra in Japan, graduated from a missionary school in the foreign settlement of Tsukiji in Tokyo before going to the Music Academy, as did the composers Koyama Sakunosuke and Kitamura Toshiharu and the organist Shimasaki Akataro. The pianist Tachibana Itoe's mother belonged to the same church as the sisters Koda Nobu and Ko, who were the first music students sent to the West.[20] To be sure, the close relationship with Christianity can be easily explained by the fact that the church was a place for contact with foreign music. In those days there were very few occasions for the Japanese to come across the new sound from the West, particularly for those in the countryside where its performance, whether in a concert or a military parade accompanied by a band, was more unusual than in the capital. The hymns sung in church were among the rare Western pieces to be heard, and the organ there one of the rare Western instruments to be seen. We often meet with the case that churchgoing provided the opportunity to get to know Western music. Shimasaki, for example, who began studying the organ in Leipzig in 1902 and taught at the Music Academy after coming back to Japan, accompanied his parents to church and got acquainted with hymns and the organ in his childhood. This led later to his decision to study at the Music Academy.[21]

It would be too simple, however, to suppose that the church merely provided the opportunity. We should rather assume that belief in Christianity and love for Western music were based on the same mentality, which worked behind the above-mentioned trend in art and literature as well, and that this resulted in the personal overlapping. Why did these Western things appeal to the »losers of the Restoration«?

III.

In the upheaval of the Restoration and the subsequent reform age, the Samurai class experienced a tremendous downfall in economic and social terms. Inflation substantially undermined their existence as vassals on fixed salaries in the 1860s. The abolishment of feudal privileges struck them; in 1876 the Meiji government forcibly reduced their stipends and converted them to bonds. Many had to seek a livelihood anew as a farmer or merchant. Their social deprivation was even more severe. The modernization of the military that had already been set in motion in various *han* at the end of the Edo era made the Samurai more and more superfluous. It shook the feudal elite's confidence and undermined their prestige in the eyes of the ordinary people. Then, in 1873, the Conscription Decree did away with their *raison d'être* as warriors. The sword, so to speak the soul of the Samurai, was prohibited in 1876. It was a natural consequence that a series of rebellions by *shizoku* occurred, culminating in the Seinan Civil War in 1877. In addition, particularly those from the former pro-Shogunate camp suffered from political alienation. The power of the Tenno's government lay in the hands of the *hanbatsu* clans that had successfully brought about the Restoration. The defeated were excluded from power or had to be content with minor positions at most.

Confronted with the manifold changes, some samurai tried in vain to return to the feudal past like desperate rebels. Some, however, sought compensation in coming to terms with the loss of orientation. In an age when no other task seemed to be more urgent than modernization, they found orientation in pursuing a thorough westernization that could be distinguished, as they saw, from that of the government. They meant to get to the intellectual basis of Western civilization. Only this would enable the building of a civilized state and the cultivation of the nation while official westernization was allegedly content with merely imitating institutions and technology. Through intellectual westernization they felt assured of psychological superiority over the Meiji regime and made up for their loss. At the same time it provided them with an alternative way of making a career. Let us take a look at Nagai Koji as a good example. He was born in 1874 in a *shizoku* family in Tottori, where the *han* had remained loyal to the Shogun in the Restoration. His parents were attracted by Christianity, and he was baptized together with them in his childhood. Nagai became interested in music as he attended services and sang hymns in church, and he decided to study at the Music Academy. His autobiography reveals some aspects of his mentality. On the eve of his departure for Tokyo he presented a poem to his parents, a poem written by Niijima Jo, a famous Christian and the founder of Doshisha University, upon his departure for studies in the United States. According to the poem: »The burning wish of youth to make their way in the outside world shall certainly be rewarded with great success.« In Nagai's view studying in Tokyo and in the West was part of the same horizon. A year later he returned to Tottori on his holidays. The provincial outlook of his hometown disgusted him so much, however, that he missed the urban landscape of partly westernized Tokyo.[22] We can clearly recognize the mentality here that was certainly shared by many of his *shizoku* colleagues: on the one hand, an unconditional devotion to Western civilization that was projected on the perspective reaching from his hometown to Tokyo with the West as the ultimate goal; on the other, the wish for personal advancement by following the alternative career path of intellectual westernization. Nagai went on to become one of Japan's leading educators and later founded a music university in Osaka.

Engagement with Western culture had thus a compensatory function for the »losers of the Restoration.« In fact, to overcome aversion this function was realized through two discourses. The first they resorted to was that of sanctifying. As we observed with Nagai, the musicians' mentality was characterized by total devotion to the West, which could easily lead to a pious attitude in exalting the West to a sublime level. In front of the holy object, worshippers have nothing to do but to kneel down; it is not a matter of individual taste. What a music journal advocated in those days is typical of the discourse. An article instructing beginners how to learn music states that the most important thing is to always bear in mind that music is a dignified art; it is a formidable mistake to regard it merely as a game or amusement. »It is out of question to expect any pleasure whatsoever from it,« the author continues. One should be absolutely strict with oneself about practicing. »Never fail to press yourself. You may not treat yourself generously for a moment.«[23] It sounds almost like a sermon that demands self-abnegation for the divinity. The journal demanded an ascetic attitude not only from the student, but also from his or her instrument. The sound it produces, as another article on a new design for the accordion states, should enhance humanity's morality. Therefore its design must be lofty, refined and, most importantly, free from all superficial ornamentation. One was, however, quite aware that the holiness of art was not always actually respected. One author deplores the impious who profane Western music: street musicians in kimonos and straw hats who sing hit songs and treat the violin like a fiddle, bands marching around the city with drums and trumpets to advertise for big stores, etc.[24] His indignation proves how deep faith in the sacredness of Western music could be. As for attending concerts, there had, of course, to be a code of behavior. First of all, one had to dress properly; anything but a frock coat or a *haori* and *hakama* is not permitted, the journal says. It is important, as well, to behave politely. One should make way for others in the aisle, for example. Discourtesy such as pointing at others in the audience with one's finger is out of the question, not to mention rising up in one's seat or leaving the hall during a performance.[25] Musicians shared in the discourse, too. Koda Nobu, for example, as she generally viewed the music situation in Europe, made particular mention of audiences' manners there. She spoke harshly of allegedly rude Westerners in Yokohama who, in comparison with the refined atmosphere in a concert hall in Vienna, applauded by stamping the floor with their feet and sticks.[26]

The second discourse to counter aversion was that of officializing, that is, exalting the engagement in Western music to an official duty. Making music was no longer the private activity of each musician but a public occupation. Whether they personally liked it or not did not count. They should rather switch off their feelings and concentrate on their duty. We can recognize the discourse of officializing most evidently in a group of professors led by Toyama Masakazu, who played a significant role outside the Music Academy in establishing Western music in Meiji Japan. All of them, having once studied abroad, mainly in the United States, were ardent westernizers; they pleaded, for example, for the introduction of the alphabet in place of Chinese characters to modernize Japanese and for a reform of poetry according to the Western model. The professors were active in music, too. They founded a club to learn Western singing, and the Music Society of Japan (*Nihon Ongakukai*) soon grew out of it. The society, the first of its kind in Japan, gave concerts regularly to refine social life, as its statute said. With a number of notables among its members and its office housed in the *Rokumeikan*, a Western-style clubhouse, the society contributed to popularizing Western music above all in the upper classes.[27] The Music Academy itself owed much to the professors; it was their public proposal to the ministry for a special music school that gave birth to the Academy.[28] While they set a high value on music in public life, they sought pleasure elsewhere in their private sphere. When they got together as good friends from student days at an elite school, they preferred visiting a kabuki theater, listening to *joruri* or amusing themselves by *kodan* or *rakugo* tellers.[29] Singing a Western melody for fun would not have occurred to them at all.

The officializing discourse was of assistance to musicians, too. It is noticeable that not a few of them had a firm grounding in traditional music: Motoori Nagayo, a composer famous for his children's songs, had professional-like skill in the *shamisen*; the hobbies of Tamura Torazo, a leading music pedagogic, were the *utai* and the *gekkin*. Even Koda Nobu, the first full-fledged Western musician who studied at the Viennese Conservatory for several years, liked playing the *koto* for fun. Her sister Koda Ko demonstrated her skill at the *koto* in an ethnological lecture in Leipzig while she was studying the violin at the Music College in Berlin.[30] We can suppose that Western music was for them, first and foremost, an assigned task to be engaged in selflessly. It did not necessarily give them emotional satisfaction. They tried to recover from the exhaustion of this official duty by means of traditional instruments during their leisure time.

The two discourses of sacralizing and officializing were actually closely connected. Serving the common supreme value was itself a public activity. Officializing Western music, in turn, could be justified by its common significance. It helped to diminish the aversion to the strange sound or even make people experience it as beautiful from the start.

IV.

We should be clear that the compensatory function of Western music needed a precondition, i.e. that the »losers of the Restoration« had a strong nationalist sentiment. They thought of themselves as competing with the Meiji government in contributing to the young nation-state. In their view, engaging with Western music coincided with the national cause. Isawa Shuji, the founder and director of the Institute of Music, is a good example. In the Music Academy, too, he wielded decisive influence as its first director. It is no exaggeration to call him, although no musician himself, the father of Western music in Meiji Japan.

Isawa, born in 1851 in a minor Samurai family in a *fudai-han* in Shinshu, had a remarkable career as a bureaucrat in the Ministry of Education. He was sent to the United States for study in 1875. During his three-year stay there he became convinced of the educational significance of music and proposed its introduction at home, which resulted in the Institute of Music.[31] Isawa was a devoted advocate of westernization. He kept close company with the modernist professors around Toyama. Their singing club owed much to Isawa's help. As its de facto chairman, he was responsible for the management of the Music Society of Japan. It was under his leadership that the Music Academy began training expert players in addition to schoolteachers. Promising students were sent abroad for further training. Upon its founding the Academy looked to Europe for a first-class artist as its teacher. The Austrian Rudolf Dittrich, a disciple of Anton Bruckner's and later a professor of organ at the Music Academy of Vienna after returning from Japan, accepted the invitation and taught there for six years starting in 1888.[32]

Isawa was also a passionate nationalist. An episode during his stay at a teacher's college in Massachusetts shows how eager he was to take an active part in building the Meiji state. In the college he had enormous difficulties in a subject. It was music; he could not sing on key in the Western scale no matter how hard he tried. The college kindly offered to exempt him from the subject. But Isawa rejected the offer indignantly, saying he would have no excuse back home for not having studied the full course and managed to master it after all.[33] Under his directorship the Institute of Music's pedagogy took on a national orientation. The songbook compiled for elementary school contained, for example, a piece about a butterfly gaily flying around in a flower garden. The seemingly innocent text actually had an indoctrinating meaning. According to the authors, the butterfly stood for the subjects enjoying themselves on the Tenno's lap, and the song was intended to convey the monarchical sentiment subliminally to the pupils.[34] In fact, Isawa later initiated a movement of patriotic educationalists. The Association for National Education (*Kokka Kyoikusha*), which he founded, aimed to »promote loyalty to the Emperor and devotion to our country.«[35] After the Japanese-Chinese War in 1894-1895, he took part in the colonization of newly obtained Taiwan by helping to give the populace a Japanese education. Interestingly, Isawa's nationalist disposition can be observed even in music. He did not entirely want to introduce Western music as such. In his view, music was a product of the climate and national character and could not be simply transplanted from one country to another. His goal was to amalgamate imported and native music into a new »national« music. He once made it clear in a letter to the American Luther Whiting Mason, the first foreign teacher at the Institute of Music, »You can clearly see that our aim is not the total adoption of European or American musics [sic], but the making of a new Japanese music.«[36] It seems that Isawa was filled with contradictions: a devoted Westernizer or a passionate nationalist? A promoter of Western music or an advocate of »national« music?

What seems at first glance to be contradictory can be resolved when we differentiate the levels of his thinking. Let us begin with the question of why it was necessary for him to rely on Western music to create »national« music. We suppose that a genuinely national character would have emerged if music had developed on its own. But this apparently did not occur to Isawa for a moment. For the superiority of Western music is for him simply too evident. He thinks it is based on scientific laws and therefore objective and rational. He refers, for example, to a physical experiment in which making the length of a string different produces tones of different height. Given the theories of temperament and harmony and mechanical instruments like the piano, which were developed on the basis of natural scientific knowledge, Western music must have universal significance, he thinks.[37] Speaking of cultural superiority, Isawa's was far from a pluralist viewpoint. In fact, he stresses that human emotions are basically the same at every age everywhere in the world.[38] He will not, to be sure, deny the influence of one's surroundings entirely; he does mention it in regard to »national« music. But compared with universal human nature, it is only of minor importance. Putting one and the same scale on human development as measured by Western values leads him inevitably to an evolutionist outlook with a Eurocentrist bias. He thinks every culture can be evaluated uniformly and aligned correspondingly along an evolutional scale. Japan, just in the middle of it, still has a way to go to catch up with the West.

Isawa thus adhered fully to Western universalism. He was completely convinced of the ultimate value of modern civilization and believed in a single evolutionary path to it. He was a nationalist in so far as he sought to give his country as high a position as possible on the scale of civilization. In other words, he was not a Japanist who, like the ultra-nationalists before World War Two, denied the universal significance of Western civilization and adhered to the allegedly absolute uniqueness of their own country. Acknowledging Western values, however, he failed to understand that each culture has its own music life. Nor could he comprehend that Japanese music is fundamentally different from that of the West. What he claimed for »national« music was, after all, just adding some native taste to the universal dish.

We should not blame him for lacking an understanding of cultural plurality for he was hardly alone in this view at that time. As for the distinctiveness of Japanese music, it was difficult even for an expert in traditional instruments to recognize it. A *koto* artist, for example, asserted, like Isawa, that the basic structure of Japanese music was entirely identical with that of the West.[39] Isawa's evolutionist view was shared by many contemporaries as well. The social evolutionism above all of Herbert Spencer, was very popular in the Meiji era.[40] In Isawa's surroundings there were many followers of the English philosopher. Mori Arinori, the Education Minister and Isawa's mentor, was one of the most influential social evolutionists in his day.[41] Takamine Hideo, an educationalist and later the director of the Music Academy, often read Spencer's works.[42] Toyama, the leader of the group of professors referred to above, devoted most of his lectures on sociology at the university to the philosopher's social darwinism.[43] But more generally, Western universalism was a kind of intellectual axiom in modern Japan. The West was the only goal for the country, and the only model to take its direction from. This was by no means surprising when we consider the overwhelming superiority of the West to the rest of the world in terms of material affluence, military power, and advanced science and technology. The Meiji Japanese, feeling sometimes threatened but more often attracted by the West, were eager to overcome the gap. Everything from the West shone with the nimbus of civilization behind it. The musicians were among those who were fascinated by it. They also breathed the intellectual atmosphere of the time, consciously or not.

V.

In this article I have pointed out that social and cultural factors played a major role in the reception of Western music in modern Japan. It might sound strange that the transfer of a sensuous thing like music does not rely completely on aesthetic factors. But this is, in a way, rather a normal mode of cultural transfer. For, when there is no benevolent attitude on the recipient's part beforehand, no cultural goods can be transferred. People usually react to alien things with rejection or, at best, indifference. So the automatism of rejection must be switched off for the transfer to proceed successfully. This cannot happen intrinsically, however, by evaluating the contents of the imported goods in question because the existing value criteria are not in the newcomers' favor. A certain bias is therefore indispensable. There must be the idea from the first that the foreign objects are in some way desirable. The bias in the case of the »losers of the Restoration« was their wish to compensate for their degradation. Also desired was a breakthrough in the socio-cultural stalemate, which was the case in many Meiji intellectuals' enthusiasm for Richard Wagner's musical drama at the turn of the century.[44] In any case, the meanings that adhere to something foreign come from its place of origin and not from the cultural goods themselves. Cultural transfer never proceeds in a vacuum but on the field of social and cultural factors. The reception of Western music in Meiji Japan illustrates this point very clearly.

NOTES

1 Osatake Takeshi, *Bakumatsu kengai shisetsu monogatari*, Tokyo 1989, 29.

2 Masumoto Masahiko, *Yokohama Gete za*, Tokyo 1986, 98.

3 »Jinsei no teido ni chui seyo,« in *Ongaku Zasshi* 18, 1892.

4 Kojima Yoshiko, *Ongaku kara mita Nihonjin*, Tokyo 1997, 29; Koizumi Fumio, *Nihon no oto*, Tokyo 1994, 306; Shibata Minao, *Nihon no oto wo kiku*, Tokyo 1987, 179. This does not mean, however, that Japanese music life was completely free of outside influence. *Shamisen*, its most important instrument, is, for example, of Chinese origin.

5 Dan Ikuma, *Watashi no Nihon ongakushi*, Tokyo 1999, 127.

6 Erwin Baelz, *Das Leben eines deutschen Arztes im erwachenden Japan*, ed. Toku Baelz, 3rd ed., Stuttgart 1937, 54; Ottmar v. Mohl, *Am japanischen Hofe*, Berlin 1904, 65f.; Bert Becker (ed.), *Georg Michaelis. Ein preußischer Jurist im Japan der Meiji-Zeit. Briefe, Tagebuchnotizen, Dokumente 1885-1889*, Munich 2001, 531.

7 For example Nomura Koichi, *Oyatoi gaikokujin – ongaku*, Tokyo 1971; Masui Keiji, *Nihon no opera*, Tokyo 1984; Nakamura Rihei, *Yogaku donyu sha no kiseki*, Tokyo 1993. The same is true of the recent work of literary historians, cf. Nakamura Kosuke, *Kindai nihon yogakushi josetsu*, Tokyo 2003.

8 Osa Shizue, »Kokka to kokugaku no iso,« in *Bakumatsu Meiji ki no kokumin kokka keisei to bunka henyo*, eds. Nishikawa Nagao/Matsumiya Shuji, Tokyo 1995.

9 Horiuchi Keizo, *Ongaku 50 nen shi*, Tokyo 1991; Nakamura Rihei, *Kirisutokyo to Nihon no yogaku*, Tokyo 1996.

10 Tsukahara Yasuko, *19 seiki no Nihon ni okeru Seiyo ongaku no juyo*, Tokyo 1993; *Tokyo Geijutsu Daigaku 100 nen shi*, vol.1, Tokyo 1887.

11 Ibid., 36ff.

12 »Ongaku gakko sotsugyo shiki,« in *Doseikai Zasshi* 6, 1897.

13 »Koto Shihan Gakko Fuzoku Ongaku Gakko sotsugyo shosho juyo shiki,« in *Ongaku Zasshi* 59, 1896, 37.

14 »Ongaku kyudaisha,« ibid., 36, 1893. This list does not, however, testify to much. Those listed are to a certain degree identical with the graduates of 1896, for the average learning period lasted for three years at that time.

15 Cf. Kamatani Shizuo, *Kohaku no fuga*, Tokyo 1998, 129ff.

16 »Ongaku fukyu hikaku hyo,« in *Ongaku Zasshi* 36, 1893, 15.

17 »Ongaku gakko sotsugyo shiki,« in *Tokyo Nichinichi Shinbun*, 26 March 1909.

18 Sato Doshin, *Meiji kokka to kindai bijutsu*, Tokyo 1999, 64ff.

19 Kimura Takeshi, »Kindai bungaku no shuppatsu ki,« in *Nihon kindai bungaku shi*, ed. Nihon Kindai Bungaku Kai, Tokyo 1966, 20.

20 Noguchi Koichi, *Ginza monogatari*, Tokyo 1997, 266; Yasuda Hiroshi, *Shoka to jujika*, Tokyo 1993, 293ff., 299; Kocho Hisako, *Taki Rentaro*, Tokyo 1968, 96; Tachibana Itoe, »Ahikawarazu no mukashibanashi de gozaimasu,« in *Doseikaiho* 216, 1935.

21 Kikuchi Tetsutaro, »Shimasaki sensei no kotodomo,« in *Doseikaiho* 194, 1933.

22 Nagai Koji, *Koshi kata 80 nen*, Osaka 1954, 28, 56.

23 »Ongaku shogakusha no kokoroe,« in *Ongaku* 10-2, 1906.

24 Takahashi Koji, »Tefukin kairyo koan,« in *Ongaku Zasshi* 43, 1894; idem, »Ongaku gukan,« ibid., 45, 1894.

25 Kimura Tadashi, »Ongaku kaijo ni okeru reisetsu ni tsuite,« ibid., 18, 1892.

26 Koda Nobu, »Ongakukai soshiki no ippan,« in *Doseikai Zasshi* 5, 1897. It would, of course, be going too far if we assumed the sanctifying discourse to be specific to Meiji Japan. It is widely known that turn-of-the-century German intellectuals had a pseudo-religious inclination toward music. Cf. Helmuth Plessner, *Die verspätete Nation: Über die politische Verführbarkeit bürgerlichen Geistes*, Stuttgart 1959, 92. The question of how much it influenced the Japanese discourse, however, must be left to future work.

27 Horiuchi, op.cit., 180; Shokakai to the Institute of Music, 4 Aug. 1885, in *Isawa Shuji Monjo*, Kamiina Kyodokan; Nakamura Rihei, *Yogaku donyusha*, p.591.

28 *Tokyo Geijutsu Daigaku 100 nen shi*, vol. 1, 285ff.

29 Mikami Sanji, *Toyama Masakazu sensei shoden*, Tokyo 1987, 19; Hozumi Shigeyuki, *Meiji ichi hogakusha no shuppatsu*, Tokyo 1988, 71, 74, 101ff.

30 Kindaichi Haruhiko, *Jugoya otsukisan – Motoori Nagayo, hito to sakuhin*, Tokyo 1983, 75ff.; Maruyama Tadaaki, *Genbun icchi shoka no soshisha Tamura Torazo no shogai*, Tokyo 1998, 72; Norisugi [Yoshihisa], »Aru kai no koto,« in *Doseikaiho*, 184, 1932; »Japanische Musik,« in *Leipziger Tageblatt*, 19 March 1902, morning issue.

31 Kaminuma Hachiro, *Isawa Shuji*, Tokyo 1962; Yamazumi Masami, *Shoka kyoiku seiritsu katei no kenkyu*, Tokyo 1967.

32 Irene Suchy, *Deutschsprachige Musiker in Japan vor 1945. Eine Fallstudie eines Kulturtransfers anhand der Rezeption abendländischer Kunstmusik in Japan*, Diss., Vienna 1992, 100ff.; Hirasawa Hiroko, *Rudolf Dittrich. Leben und Werk*, Diss., Vienna 1996.

33 *Rakuseki Isawa Shuji sensei*, Tokyo 1919, 36.

34 *Tokyo Geijutsu Daigaku 100 nen shi*, vol. 1, 208.

35 Isawa Shuji, *Rakuseki jiden*, Tokyo 1912, 182.

36 Isawa to Mason, 1 July 1889, in *Isawa Shuji senshu*, Tokyo 1958, 967.

37 Isawa Shuji, *Yogaku kotohajime*, Tokyo 1971, 6; Megata Tanetaro, »Waga kogaku ni shoka no ka wo okosubeki shikata ni tsuki watakushi no mikomi, 1878,« in *Tokyo Geijutsu Daigaku 100 nen shi*, vol. 1, 16. This memorandum was published under the name of Megata, Isawa's superior. But it is confirmed that Isawa wrote it.

38 Isawa, *Yogaku kotohajime*, 21.

39 Ibid., 47.

40 Yamashita Juichi, *Supensa to Nihon kindai*, Tokyo 1983.

41 Inuzuka Takaaki, *Mori Arinori*, Tokyo 1986.

42 *Takamine Hideo sensei den*, Tokyo 1980, 43.

43 Mikami, op.cit., 33.

44 Takenaka Toru, »Der Wagner-Boom in Meiji-Japan,« *Archiv für Musikwissenschaft*, 62-1, 2005.

Paying the Piper:

Mychael Danna's Music for *The Sweet Hereafter*

SIMON WOOD

When Dorothy and Toto are swept away from Kansas to the Land of Oz, they are literally plucked out of one cultural state and dropped into another. Against their will, they are carried off from the comforts of »point a« to the strange and mysterious »point b,« the opposite pole of the magnet – a sphere which is simultaneously excessive in both its natural and unnatural conditions. Dorothy is eventually offered the choice: to stay where she is or to return to the familiarity of »point a.« There is no opportunity for resistance or for the enactment of radical change – there is no »point c.«

I would like to begin with the assertion that much of the music produced for the Hollywood model of the narrative film is designed to operate within a framework of binaries such as self/other, male/female, technology/nature, or individual/collective. Like Dorothy, the Hollywood composer may move from one sphere to another, but the two must remain at odds with one another in what Royal S. Brown describes as the »hierarchically weighted dialectics« of modernist practice (Brown 236). At a purely pragmatic level, this makes a great deal of sense. As it is the intention of a film's producers that the viewer accepts the narrative as reality (requiring a suspension of disbelief which allows the viewer to engage with the narrative), it is no surprise that most of the codes, conventions and strategies of narrative film are designed to minimize the constructed nature of the medium. Music assists this process by helping to guide the viewer through the film, accentuating that which is important while reducing undesirable ambiguity in the viewer's interpretation of events through the limiting of possible readings.

Certainly one of the most utilized conventions in film scoring practice is the incorporation of the »exotic«: the deployment of musical structures and instrumental timbres derived (with various degrees of accuracy) from the sounds of non-Western cultures. Like the Land of Oz, these cultures are characterized by excess and absence: often an excess of sexuality and a lack of restraint. Historically, this practice generally results in a colonialist parody: complex musical traditions are grossly simplified and stripped of their social context – transformed into signs that, through repetition, become an easily decoded shorthand for the culturally competent spectator, which is to say, someone who has seen numerous Hollywood films. Examples of this practice abound: from Max Steiner's tribal drums in *King Kong* (1933) to John Williams' use of diminished scales orchestrated for double reeds accompanying Indiana Jones' arrival in Cairo (*Raiders of the Lost Ark*, 1981). The sounds are reduced to a set of commodities to be deployed and consumed, uncontested by the spectator whose desire is satisfied only by the certainty of linear narrative.

However, recent years have seen the rise of »world music« as an increasingly popular genre of music within the West. The term is certainly problematic as it simply introduces a new binary, while being so vague as to be rendered almost meaningless. One can also suggest that »world music« is simply another form of »orientalism« in which Western ears reduce the sounds of other cultures to a kind of fashion, once again converting cultures into capital to be consumed without any understanding of origins or context. However, the West's indulgence in world music has brought about a generation of musicians and composers who claim to engage with non-Western cultures in a respectful and informed manner. Within the genre of film music, such an approach offers an alternative to the established conventions, to subvert the constructed sense of opposition and to introduce a higher level of ambiguity into the reading of a narrative, allowing space for a variety of viewer interpretations and, thus, reworking the binary into a continuum.

Such is the claim of Canadian composer Mychael Danna, who is perhaps best known for his work with Canadian director Atom Egoyan. Danna states that he had no intention of becoming a film composer and has little time for the conventional approach: »[T]he John Williams/Jerry Goldsmith school of film music [makes] me vomit« (Morgan 245). Since his 1987 feature debut in Atom Egoyan's *Family Viewing*, Danna has consistently incorporated his interest in Middle Eastern and South Asian music and the music of the European medieval and Renaissance periods with electronics and minimalist techniques to create music which Egoyan describes (perhaps problematically) as: »transcendent, otherworldly, yet rooted in the very real emotions that define a community.« Note that Egoyan's use of the word »yet« suggests that, at least in his opinion, Danna's music draws the points of opposition together: inside/outside, transcendent/commonplace, self/other. Danna explains that his approach to the use of non-Western instrumentation is informed by some level of understanding of the traditions responsible for the production of these sounds:

> I've always been annoyed by the kind of patronizing and ignorant film-scoring uses of these musics. When you see a film that's about Middle Eastern terrorists and you hear tablas and sitars, things like that really get on my nerves. It's just ignorance. It's something that I wanted to make sure that I wasn't doing. You can do that if you have a reason, but if your reason is that you just don't know any better, that's not a good reason. I think, having some knowledge and some understanding of those cultures, what the music means to them, and how it's used is important. It doesn't mean that you have to follow that all the time, but I think that you should be knowledgeable. If you're going to stretch the boundaries or bend the rules, you should know that you are doing that. As opposed to doing it by accident. (Adams)

Certainly, Danna presents himself as someone who offers an alternative to the status quo. However, one could argue that Danna's link between knowledge of a culture and the use of its sound places him in a privileged position as one who can speak for the absent other. His stress on the value of informed intent, as opposed to an uninformed pastiche, suggests that his approach may be informed by a romanticized concept of an »authentic« cultural use. To see how Danna's approach to the use of non-Western (or non-conventionalized) music within a narrative film actually works, I turn now to a discussion of *The Sweet Hereafter*.

Based on Russell Banks' novel, Egoyan's 1997 film *The Sweet Hereafter* relates the tale of a small town in which all of the children but one are killed in a horrific accident involving the local school bus. Mitchell Stephens, a lawyer who is losing his own child to drug abuse and Aids, attempts to persuade the townsfolk that they are entitled to take legal action against the maker of the school bus. He convinces some of the grieving parents to relive the tragedy and, in the process, they come face to face with the disappointments, transgressions, and limitations of their own lives. The film moves back and forth through several distinct timelines: moments in the life of the town before the accident; the morning of the accident; Stephens' arrival and his gathering of the families for the lawsuit; a plane flight in which Stephens finds himself seated next to a childhood-friend of his daughter Zoe, two years after the accident; and a flashback in which Stephens relates the story of how as a baby, Zoe almost died from an allergic reaction. Woven throughout the film is the story of *The Pied Piper of Hamelin*, in which all of the children but one are spirited away as retribution for the wrongs of their parents. Although not a part of Banks' original story, Egoyan felt that the story, which he knew from Robert Browning's poem, was »an amazing controlling metaphor« that offered a comment on the narrative (Porton). The director states that:

> [w]hat's remarkable about that poem is not only the idea of a town that loses its children, but also this idea of reward and punishment, which is the darker side of this whole story, people looking for a way to deal with this loss through a reward. Something as crass as a financial settlement might give a sense of closure to what is otherwise unbearable. But of course, it's never that simple, and that's one of the ideas that the film presents. (Fuchs)

It was this parallel narrative-within-a-narrative which served as the inspiration for Danna's choice of instrumentation which features the Toronto Consort, an early music ensemble, augmented with the use of the Ney, an Arabic wind instrument. According to Danna:

> There's a terrible tragedy that happens and a lawyer from outside comes and tries to seduce the people into a big class action lawsuit. [...] The Pied Piper story, as you know, takes place in a medieval, tiny town. Then the exotic, nasty, Middle Eastern piper [ar-

> rives] – [note that the use of »exotic« and »nasty« is ironic]. So, really that was the choice of the music right there. It was early music instruments on one hand, and then an Eastern Ney from Iran. (Adams)

In Danna's view, the use of period instrumentation in association with the small, isolated town evokes the sense of an idealized past: a simple community more in tune with the rhythms of nature, yet naïve, unable to resist the lure of the urban lawyer/exotic piper (who must eventually be paid). It may be seen as an example of what Caryl Flinn describes as »the idea of looking backwards, of raiding history for utopian models of the future« (Flinn 125). Although Flinn is most concerned with ways in which the ideology of romanticism permeated the music of Hollywood's »classical studio« era of the 1930s to the 1950s, Danna employs a similar strategy through the evocation of a pre-industrial society, as a parallel for a contemporary setting, describing the music as being »suggestive of another time and another place« (Adams).

However, if we consider the associations created by the intersection of music and visual, Danna's view of the function of the music immediately becomes problematic. The opening credits are accompanied by the Consort. The opening theme is in triplet time, its tempo slow and stately. Scored for lute and recorder the theme most closely resembles the courtly dance music of sixteenth-century Europe, which technically makes the influence one of the Renaissance. (This presents a small inconstancy, as Danna often uses the term medieval in his discussion of the music.) The visual presents us with a slow track across a wooden surface, gradually revealed as a floor upon which we see a young couple asleep, an infant child between them. We later come to realize that this is the lawyer and his family, quietly resting in a utopian past, accompanied by the music of the Consort. The visual then moves to the interior of a car. The viewer is placed in the back seat, looking over the shoulder of the driver through the windshield, where we see that the car is passing through a carwash. Musically, Danna continues to use the main theme but now arranges it for the full ensemble including percussion. The change of scene and music jarringly raises the level of intensity, while the slow, relentless motion through the carwash gives the scene a feeling of a procession.

The credits come to an end as Stephens receives a call from his daughter Zoe on his cell phone. Zoe is shown at a payphone in a rundown urban industrial area, the sound of trains and police sirens in the background. It is night, and the florescent lighting of the phone booth gives her a pale, deathly complexion further emphasized by her garish makeup. The conversation quickly establishes that Zoe is estranged from her father and has become a drug addict. Stephens is clearly torn by his desire to help his daughter and the feeling that she may be beyond hope. After an uncomfortable exchange, the clearly irrational Zoe hangs up on her father. At this point, Stephens realizes that his car has become stuck in the automated carwash, as if the call from his daughter is a stain that cannot be removed by technology. Although the carwash promises a renewal, a cleansing of the exterior, it comes at the cost of surrendering control. The driver cannot steer or break, producing a feeling of vulnerability. And at this most vulnerable moment, the technology of the carwash fails him. His attempt to use the cell phone to call for help is also thwarted, as the operator cannot hear him clearly. Indeed, the cell phone, a signifier of status, of communication, of a mastery of technology for communication becomes a constant interruption for Stephens. Each of his conversations with the grieving parents is interrupted by a call from Zoe, a reminder that he has failed to save his own child.

All of the music that accompanies the conversation is diegetic. Zoe's urban nightmare is scored with the sound of a drum machine emanating from a nearby car stereo. Its insistent mechanical groove dulled to a bass-heavy throb by the windows of the phone booth further enhances her sense of isolation and distance. Stephens is listening to the sound of the Consort, its presence perhaps representing the lawyer's longing for that last perfect moment with his family, asleep on the floor of a cottage. It is also a strong contrast with the technology of Zoe's drum machine and the motorized thud of the carwash itself. Stephens is depicted as a man of the past; his connections with the present and his interactions with its technological sphere are doomed to failure. Indeed, this is an interpretation in keeping with Danna's link between the Consort and the utopian past. But Stephens is not the piper. Rather, his love of the past and his inability to cope with the present suggest that he is the one who has been duped by the exotic other. He is the one whose perfect past has been stripped away from him by the inexorable passage of time.

Once he escapes from the carwash, Stephens wanders through the deserted gas station. More diegetic music accompanies his exploration. This time, the source is an electric guitar resting on its amp. The volume is up, and the strings resonate with feedback, their timbre overtone rich with distortion. It is abandoned technology, a voice without a soul, the endless mechanical cycle awaiting the touch of a human hand. Stephens reaches out to it, but the sound continues as if the lawyer is unable to silence its keening wail, helpless to calm the technology. At this point, the music moves back to the non-diegetic frame. The textures thicken with the addition of synthetic drones and, for the first time, the sound of the Ney is processed with the addition of reverb and echo, just as Mitchell sees the wreckage of the school bus. The sound of the »exotic« merges with the ghostly screams of the absent children, but the scene shifts back in time as we see the bus at the local fair, the screams of terror becoming cries of delight. It is as if the Ney is a siren's song, calling out to Stephens that here is a chance for reclamation. The bus and the lives that were lost within present an opportunity to make good on the failings of his own life, or, as he states to the grieving parents of Bear Otto, »to give anger a voice« (Egoyan).

On the one hand, Stephens *is* the urbanized, exotic outsider. He is the one who claims to understand the world and who can act for the naive townsfolk; he is the one who, in his own words, »know[s] better.« The lure of the sophisticated, cultured, urban piper, his sound rich in timbre with its technological manipulation, would seem to be a strong match for Stephens' role. And yet, as we have seen, this is the contradiction of Stephens' character. For if the Consort represents the utopian past, the idyllic memory of a pre-technological era, its connections with Stephens would suggest that the lawyer longs for a return to the past. Mitchell is not the piper but simply one of the townsfolk, outraged that his daughter has been spirited away by forces beyond his control and understanding. Indeed, in a moment of terrible insight, Mitchell confesses to Billy Ansel that »we've all lost our children. [...] They're dead to us. They kill each other in the streets. They wander comatose in shopping malls. They're paralyzed in front of televisions. Something terrible has happened that's taken our children away. It's too late. They're gone.« Through the various musical connections the film makes with the lawyer, Danna and Egoyan have undermined any reading that definitively positions Mitchell as the piper.

Closer to the story of the Pied Piper is the incestuous relationship between Sam Burnell and his sixteen-year-old daughter, Nicole. Indeed, it is Nicole who is directly associated with the poem, which is first heard as she reads it to the Ansel twins, Jessica and Billy, the night before the accident. Later, at key narrative moments, she is heard in voiceover, reciting specific verses. Nicole is introduced as a singer practicing at the local fair, her band a motley crew of cello, drums, harmonium, and guitars. Behind the band is a poster of a Janus-like Nicole, the two identical faces looking in opposite directions, a symbolic depiction of Nicole's plight. The only audience for the performance is Sam. Dressed in overalls, with a tool-belt slung around his waist, he is positioned as the architect, the workman who labors behind the scenes to make his daughter's musical dream come true. After the performance, Nicole's desire for her father's approval is obvious. She asks him his opinion several times, questioning his sincerity until he replies: »Really. It was awesome.« She throws her arms around him and cries, »I'm so happy, daddy.« He leads her away for ice cream, past a pair of farmers leading their cow to the livestock pens. The connection between the two is not coincidental, as the warm supportive relationship between father and daughter is a chimera covering a much more disturbing reality.

After babysitting for Billy Ansel (who is having an affair with his neighbor, Risa Walker) on the night before the accident, Nicole is taken to an abandoned barn by her father. Wearing a short tight dress, she presents a Lolita-like figure of transition: a child on the verge of adulthood. Like her depiction on the band's poster, she looks in opposite directions simultaneously. She follows Sam to the doorway but pauses at the dark entrance to the barn, her expression a mixture of affection, confusion and resentment. Sam, a guitar in his hand, (an allusion to the piper), looks back at his daughter, his eyes betraying guilt and fear. He turns to the barn and leads her into the darkness. Over this scene is the sound of the Ney, the »otherness« of its timbre accentuated with echo and reverb, and the voice of Nicole reciting the verse in which the children of Hamelin are taken from their homes: »When, lo, as they reached the mountain-side / A wondrous portal opened wide / As if a cavern was suddenly hollowed; And the Piper advanced and the children followed / And when all were in to the very last / The door in the mountain-side shut fast.« At the end of the scene, we see Sam and Nicole deep inside the barn, surrounded by bails of hay, illuminated by candles. As the camera pans upwards, Sam slowly pulls his daughter to him and kisses her.

In this scene, the sound of the Ney is linked to the transgression of normative behavior and the risk of the corruption to which such behavior can only lead. Its plaintive call is a lure for the innocent, drawing them into danger with the promise of something beautiful, something exotic that cannot be realized in this world. And yet, if this is the case, who is the piper? There is clearly a literal correlation, as Sam the piper leads Nicole into the darkness of the barn. But unlike the children of Hamelin, Nicole is not fooled by the piper's magic. She sees through the guise but follows out of love and perhaps even pity. And if Sam is the piper, what is the wrong for which he seeks retribution? Who is to be punished by this action? He is divorced, and the narrative suggests that he may be an ex-musician whose dreams of success have faded. We might posit that his actions are to make amends for this loss in some way, but only Nicole is aware of the punishment. Furthermore, while she is recovering in hospital, claiming that she has no memory of the accident, Sam appears to offer protection when he urges that she should not »even try remembering« for he knows that if exposed, his actions would bring about his own punishment at the hands of the community. There is also no apparent connection between Sam's status as the piper and the bus accident. Indeed, Nicole is the only child to survive the accident, suggesting that Sam is to be spared the loss faced by the other parents of the town.

However, like the lame boy who was left behind, Nicole is now confined to a wheelchair by her injuries, paralyzed from the waist down. Her injuries have imparted a change in Nicole and, like the two faces on the poster that frames her band's performance, we see a different side of her following the crash. Externally, she appears unaltered, but she now looks in a different direction. The night before Nicole is to testify that Dolores, the bus driver, was adhering to the speed limit at the time of the accident, the Burnells are visited by Billy Ansel.

Despite the loss of his twins, Ansel has refused to join the lawsuit. He asks Sam and his estranged wife to withdraw from the legal action, and when they refuse, he questions their motives suggesting that the pursuit of money is damaging the community as much as the loss of the children. Sam, clearly shaken by the encounter, checks on his daughter, asking her if she is ready to testify the following morning. But Nicole is no longer the naïve, trusting daughter. She understands what her father has done to her. In her suffering she has found wisdom and courage. When he asks her why she is so quiet, she pointedly answers: »We didn't use to have to talk a lot, did we, daddy?« Perplex and growing more nervous, he asks what she means. Nicole replies: »I mean, I'm a wheelchair girl now, and it's hard to pretend that I'm a beautiful rock star. Remember, daddy? That beautiful stage that you were going to build for me? You were going to light it with nothing but candles.« His eyes wide with fear and guilt, Sam is rendered speechless. Eventually he stammers out the time they will depart the following morning and leaves the room.

Musically, the scene is scored not with the ominous sound of the Ney but with The Tragically Hip's song »Courage,« heard through a portable cassette player on Nicole's bed. The song was heard earlier in the film, performed in a soft, dream-like manner by Nicole and her band, but this time it is the hard edge of the original version. Paradoxically, the silencing of her old voice by the accident signals her realization of the truth, and her decision to punish, not the town, but the piper. The song signifies both the passing of her childhood and the resolution to make something anew. Increasing in volume, it moves to the non-diegetic realm as the scene shifts to the following morning. The drive to the community center, where Nicole will give her deposition, is scored with the final verse of the song, the vocal performance and accompaniment building in intensity: *So there's no simple explanation / for anything important any of us do / and yet the human tragedy / consists in the necessity / of living with the consequences / under pressure, under pressure* (The Tragically Hip). The arrival at the community center is synchronized to the chorus of the song: *Courage, my word, it didn't come, it doesn't matter / Courage, it couldn't come at a worse time* (ibid). As the song moves into the guitar solo, we once again hear Nicole reciting the verse in which the children of Hamelin disappear into the mountain. Only this time, the music suggests that it is she who is in control, as the success of the lawsuit hangs on her deposition. The Ney is absent; the piper is silenced.

As Nicole begins her testimony, she seems uncertain. But then she claims to recall details of the day. When asked the speed of the bus, she looks at her father, and we hear in voiceover: »It's dull in our town since my playmates left / I can't forget that I'm bereft / of all the pleasant sights they see / which the piper also promised me.« Under the verse, the sound of the Ney enters, now an evocation of the abuse and a signifier of Nicole's loss of innocence. Yet this loss of innocence is what gives her the power of the piper and, at this moment, she extracts her price. She lies, claiming that Delores was speeding, effectively killing the lawsuit and her father's quest for money. From Sam's point of view, Nicole's courage couldn't have come at a worse time. The Ney hangs over the scene, as Nicole's voiceover continues with a final verse added by Egoyan: »And why I lied, he only knew / but from my lie, this did come true / Those lips from which he drew his tune / were frozen as a winter moon.« Once again, the film and its music seem to position Sam as the piper, but this time the Ney is a sign that Nicole has usurped the magic of the father to deceive, and that she is now the character that all must follow. Her testimony thwarts the plans of all those around her and, even though they know she is lying, they are powerless to intervene. In a twist on Browning's story, the lame child punishes the piper, not for being left behind but for making promises that he could not keep, for stealing her childhood. And with the lie, she reveals herself to him as one who sees through his trickery and treachery. She silences the piper, ensuring that he will never have such power over her again.

In explaining the link between the film, the poem, and the music, Danna suggests that »somebody from Iran would have been an exotic Pied Piper to a European in the Middle Ages« (Morgan 249). There may be some truth in this statement, but we have seen that the use of the Ney is not linked to that which is simply »exotic« but rather to something that transgresses what would be considered acceptable behavior for the vast majority of the film's viewers. Regardless of how informed Danna's use of the instrument may be, the elements of the film involving Sam and Nicole clearly associate the Ney with incest, child abuse, and then with the power of deception used to punish those who would transgress the social order. The source of danger is coded as the non-Western other. It is only when Nicole is able to free herself from the abuse and punish her father for his wrongdoing that we hear the sound of The Tragically Hip – the sound of the Western self signaling the (re)establishment of an acceptable social order. Such an interpretation certainly calls into question Danna's claim that he is avoiding the conventional pitfalls of non-Western musical gestures in film music. The Ney in particular seems linked to narrative elements that carry the excesses of the clichéd other, excessive in sexuality and deception, and lacking the appropriate restraint.

The point at which the two primary musical elements of the Ney and the Consort are brought together is during the scene depicting the bus crash, which is not revealed until half way through the film. Egoyan cleverly holds back the details of the accident, building the sense of anticipation and dread in viewers, as they struggle with the knowledge that the children will die but that the manner of their death remains unspoken. Our initial views of the bus interior are scored with a gentle statement of the film's opening theme heard in the recorder with a pedal ostinato in the lute. This theme clearly articulates a sense of loss and mourning, seemingly at odds with the scene of the children on the bus, yet in keeping with how this scene is revealed. The camera/spectator slowly drifts up the aisle of the bus like a spirit, capable only of observation, unable to intervene in the approaching tragedy. In fact it is the empty aisle itself, and the silhouettes of the Ansel twins at the back of the bus, waving farewell to their father, which is the object of the cinematic gaze. The rest of the children, vibrant and dynamic in their final moment of life, are pushed to the margins of the screen. The circular rhythm, both in the recorder and its accompaniment and static harmony with its pedal point, further serve to slow the pace of the scene. In fact, after I saw the film for the first time, I was convinced that the children had been shown in slow-motion. An example of what Royal S. Brown refers to as film music's ability »to imagify time,« it was only after I reviewed the film a second time that I realized my mistake (Brown 245).

As the view shifts to an external shot of the bus, the musical intensity is increased through the addition of instruments (primarily percussion) as well as the doubling of the melody by a second recorder up one octave. The Ney also becomes a dominant musical force at this point, its plaintive and frantic articulations growing louder and more insistent as the bus approaches its inevitable fate. The Consort and the Ney seem to struggle for control of the film's musical space, a struggle that is resolved only at the moment the bus strikes the guardrail. At this point, the instruments of the Consort are silenced, leaving only the sound of the Ney. Once again its »exotic« timbre is enhanced through the use of studio technology. The instrument is overdubbed several times and heavily processed through the addition of reverb, as the bus careens down the hill and out onto the surface of a frozen lake. For a brief moment, it seems that all may survive, but the screams of the frightened children are quickly dwarfed by the sounds of the ice shattering under the weight of the bus – a terrible thunder echoing across the valley and blending with the echoes of the Ney. Like the children of Hamelin disappearing into the side of a mountain, the bus slides beneath the ice, and the frozen lake claims the children of the town. The view of the accident is from a great distance, the viewer completely helpless to intervene.

Musically, the struggle for control parallels a futile yearning for the past and the loss of innocence, not only the innocence of the pre-industrial utopia but the innocence of childhood itself. Often accompanied by the music of the Consort, the time before the accident is recalled by the town's residents as a distant and now unobtainable past; the death of the children is portrayed as a pageant. Yet, as we have seen, this historicism of simplicity and naïveté is undermined by the town's dark secrets of infidelity and incest revealed throughout the course of the film. This is not an idealized paradigm for community, but a town rife with the problems of the human condition, which the residents have been unable to escape through their voluntary isolation in the mountains. We might even posit that their seclusion has contributed to this erosion of morality. The loss of their children may be read as the cost of this inability to overcome these conflicts. Like a retribution of biblical proportions, the bus accident has literally robbed the town of its future. Rather than working together to clarify a possible interpretation, the music of the Consort and the narrative seem to operate in a mutually subversive manner, the idealism symbolized by the one contrasting with the stark reality of the other.

Once again, the use of the Ney creates problems for the viewer seeking an unambiguous reading of the text. As the bus winds along the mountain road, the sound of the Ney hangs over the Consort, as the audience's knowledge of the accident hangs over the images of the children. It is a threat, both to the children and to the uneasy stability of the town. It is the mysterious piper, come to claim his price. But the threat is contradictory. Is it the exotic of the non-Western or the urban technology that takes away the children? Is it punishment for the town's corruption or their isolation? Like the music of the consort, the sound of the Ney offers alternatives to the reading of the scene. However, Danna once again seems to be resorting to the conventional approach with the use of the Ney to accompany the death of the children. The association between distorted timbre and lack of clear Western tonality with moments of threat, loss, or instability is a common feature of the Hollywood soundtrack.

So does Danna's approach really create an alternative to the more conformist strategies of Hollywood that he claims to detest? At first glance, Danna's association of the Ney with the exotic/urban piper/lawyer seems to make sense. He is the sophisticated outsider. This is subtly clarified through the consistent use of the overdubbing and electronic manipulation of the Ney, which paradoxically enhances its distance from Western timbres while placing it in the realm of Western technology. However, Stephens' difficult relationship with his daughter, manifest through his inability to deal with the tools of technology position him as simply another parent trying to deal with the loss of a child. Furthermore, the Ney is also associated with the incestuous relationship between Nicole and her father and the power struggle that emerges between them following the accident. This blurring of musical and narrative associations *do* open spaces of interpretive ambiguity, which, even at the film's conclusion, remains unresolved. The spectator cannot depend on the music to focus narrative interpretations. In fact, the music seems to undermine any attempt at making a definitive reading of the story.

One might conclude that by creating a sense of ambiguity, Danna circumvents the problematic dichotomy of film-score accompaniment, the opposition of parallel and counterpoint: either constructing a musical framework parallel to that of the narrative, creating a sense of unity within the materials of the film or constructing a contrapuntal framework and thereby exposing the artificial, heterogeneous nature of film. Parallelism is thought to be Hollywood's normative practice, while contrapuntalism is more often associated with the self-reflexive schools of European and Soviet filmmaking (see Gorbman). Danna's music seems to do both, and therefore any reading that limits its interpretation to one pole of the dichotomy cannot help but undermine itself. Instead, spectators will find their reading of the text located somewhere in between the extremes of parallelism and counterpoint. Rather than being guided through the narrative with a series of binary choices, the intention of the film becomes far more ambiguous and, therefore, the viewer becomes a far more active participant in the construction of the film's meaning. This would certainly be in keeping with Danna's views on conventional scoring practice.

However, the use of The Tragically Hip's »Courage« offers an alternative reading of the score. Although claiming to circumvent conventional approaches, Danna has clearly established a musical conflict between East and West, which functions within the film on a number of levels, such as rural/urban, nature/culture, and past/present. Furthermore, a reading guided by the established conventions of scoring practice would tend to code both sides of the musical divide – the medieval/renaissance and the Ney – as other to an unspoken self, usually articulated by a Western orchestra working within late-romantic tonality, or, in this case, a Canadian rock band. Indeed, two key moments of the film are scored not with the Consort or the Ney, but with piano, flute, and synthesized strings. The first is when Billy Ansel identifies the bodies of his children, his face a mask of shock and disbelief. The second is in the film's final moments in which Nicole's voiceover speaks of the world after the accident: »the sweet hereafter,« in which »everything is strange and new.« In both scenes, Danna abandons the exotic others: the Consort's utopian past unblemished by sin and the Ney's false promise of an unobtainable future in which all sins are forgiven. Instead, he brings the audience home with the sound of the »real« articulated through the sound of conventional, contemporary Western instruments. We are brought face to face with the permanence of our past and the imperfections of our present, in which we must deal with our mistakes as best we can, realizing that we will surely make more tomorrow. But this realization is couched in terms of a system of a self defined by the sounds of Western contemporary music, while the other is signified by the historical sound of the Consort and the »ethnic« timbre of the Ney. Although Danna's music does bring an effective level of ambiguity to the story, the Hollywood model is all too present, lurking just below the surface of this intriguing score.

Mychal Danna's use of non-Western instrumentation and musical styles in *The Sweet Hereafter* clearly demonstrates that going beyond Hollywood's accepted practice of film scoring, in regard to the use of the exotic, offers a dramatically expanded potential for the role of music in the process of narrative interpretation. It presents us with a world that is indeed »strange and new« and suggests that the difference between Oz and Kansas may be less extreme than we have been led to believe: not that there is »no place like home« but, perhaps, that there is »no place quite so *strange and new* as home.« However, even here, the conventions of Hollywood are all too present posing one last question:

No matter where we travel, can we ever really leave home...?

REFERENCES

Adams, Douglas. »An Interview with Mychael Danna.« *Film Score Monthly* 1998.

Brown, Royal S. *Overtones and Undertones: Reading Film Music.* Berkeley: University of California Press, 1994.

Egoyan, Atom. *The Sweet Hereafter.* (Screenplay). Ego Film Arts, 1997.

Flinn, Caryl. *Strains of Utopia: Gender, Nostalgia, and Hollywood Film Music.* Minneapolis: University of Minneapolis Press, 1992.

Fuchs, Cindy. »Staying Open.« http://citypaper.net/articles/122597/movies.interview.shtml Accessed on 1 May 2005.

Gorbman, Claudia. *Unheard Melodies.* Bloomington: Indiana University Press, 1987.

Kassabian, Anahid. *Hearing Film: Tracking Identifications in Contemporary Hollywood Film Music.* New York: Routledge, 2001.

Kaufman, Anthony. »An Interview with Atom Egoyan, director of *The Sweet Hereafter.*« http://www.indiewire.com/people/int_Egoyan_Atom_1_9711_21.html. Accessed on 30 April 2005.

Morgan, David. *Knowing the Score*. New York: HarperCollins, 2000.

Porton, Richard. »Family Romances.« http://www2.cruzio.com/~akreyche/frintro.html#topfri. Accessed on 1 May 2005.

The Sweet Hereafter. Dir. Atom Egoyan. Fine Line, 1997.

Tragically Hip, The. »Courage.« *Fully Completely*. Compact Disk. Sire/Wea, 1993.

Welcome to Class Europa: European Virtuosity and Global Virtuality in Mamoru Oshii's *Avalon*

MARKUS REISENLEITNER

In 1999 the Wachowski brothers released their worldwide sleeper hit *The Matrix*, which successfully combined a video game culture aesthetic with Asian animation and martial-arts style, sprinkled with a fair dosage of pseudo-Zen fortune-cookie babble on questions of postmodern simulacra, space implosion and the ontological status of reality mediated by technology. This heady gumbo was by no means the first movie to engage with such issues. Inspired by the genre conventions of cyberpunk, acclaimed Japanese anime director Mamoru Oshii had explored similar issues in his previous work, most notably in his 1995 *Ghost in the Shell*, a »formative influence on *The Matrix*« (Ruh 165), in its 2004 sequel *Ghost in the Shell 2: Innocence*, and in his ten-years-in-the-making feature film *Avalon*, which was released in 2001 in Europe and Asia. For the North American market, Miramax acquired the distribution rights at Cannes 2001 but never released it in theaters (Hoberman); however, *Avalon* still managed to spawn its own worldwide cult audience, including fans in North America (presumably returning a healthy profit for Miramax) through its DVD version, which featured, in the tradition of the paradigmatic *Blade Runner* (1982, dir. Ridley Scott), additional voice-overs to clarify some of the more opaque plot elements. Similar to Oshii's earlier animations, *Avalon* is indebted to computer game culture as much as to noir mysteries and their postmodern successor genre, cyberpunk. What is exceptional about Oshii's first »live-action« feature is that *Avalon* was filmed entirely in Poland and engages with one of the most pervasive European mythologies, the Arthurian legends. In this essay I am tracing the specific imaginings of Europe in *Avalon*, particularly the traces of history and memory on which the film literally builds its computer-graphics generated spaces.

Not unlike *The Matrix*, *Avalon* immerses the viewer in the monochromatic world of a first-person shooter game. A fairly long text prologue sets the story in the »near future, in a country in ruins,« a bleak urban environment in which the only distraction many people engage in addictively is the illicit virtual reality war game Avalon, whose top players can make a living, even a handsome profit, from the game, but they also expose themselves to the danger of becoming brain-dead vegetables (»unreturned«) when they fail to press »reset« in dangerous situations in VR.

The beginning of the movie immerses the audience into this game world, a sepia-colored battlefield populated by World War II tanks, panicked masses and a rocket-launcher-wielding warrior queen. With these shots, Oshii demonstrates from the start (like the Wachowski brothers with their widely acclaimed rotating camera effects) that he is fully in command of his visual effects: freeze frames and spectacular vector graphics of dissolving bodies being »fragged« by the formidable Ash (Malgorzata Foremniak) set the tone for a visually stunning game universe that, like many science-fiction sets, is equally futuristic and nostalgic (Sobchack 229).

The main plot pattern revolves around Ash, a solo star player who has become so adept at making a living from her exploits in cyberspace that she can cook gourmet meals for her dog with »real vegetables« – a rare thing in the decrepit real world, whose population is poor and starving, almost immobile against the backdrop of the crumbling buildings of a vaguely Eastern European cityscape. When Ash hears rumors about a secret top level of the game (Avalon, we are led to assume, works somewhat like a platform game) called »Class Special A« (she is already proficient in Class A, the highest level accessible to normal players), she becomes ambitious and starts gathering information about this place. We learn that one of her former shooting buddies, Murphy a.k.a. »Wizard,« now wasting away as a brain-dead vegetable in a hospice, is actually still somehow present and sentient in the cyberuniverse of this class. In order to gain access, Ash sets out to solve riddles based on mystical references to Arthurian mythology with the help of real books, which, in the absence of information accessible online, provide the »cheats« for her riddle-solving quest. She is warned that there is no return from this level and that everybody who has tried to gain access has ended up like Wizard. Ash forms a team with the mysterious Bishop, a »Thief« and a »Mage« (recognizable characters in the vaguely Tolkien-medieval universe of role-playing games). Not unsurprisingly, Ash manages to both solve the riddles and shoot her way into level Special A, where she meets (and has to kill) her brain-dead buddy while listening to a Japanese version of classical music playing in the background, which is being performed diegetically in a concert hall in the restored central part of Warsaw.

The »central metaphysics« of the film's plot (is the imagined/engineered world more real/compelling than the actual one?) can certainly be described as »old wine, but they're persuasively poured into smart new bottles« (Leggat 79). The tag-line of the movie (featured prominently on the website) is »this is my reality,« and the movie has been described as a »retro-futuristic game« (http://www.avalon-movie.com/index.asp). The heroine is a cross between Lara Croft and Trinity, with the same dark-haired, shapely cardboard personality (or lack thereof) and dominant yet unthreatening sexuality. The plot is relatively straightforward, like those of the computer games the script is taking its cues from, and the philosophy (as in so many Japanese anime) – garbled Zen. In these respects *Avalon* readily continues the director's earlier work and inserts itself into the canon of cyberpunk-inspired films »with its combination of high tech and old grime« (Ruh 172). The director himself claims that this pseudo-metaphysical aspect should be seen as the major focus of his movie:

> These days, many movies use the game as a subject matter, but they have nothing to do with the movie I imagine. I don't think it's a question of bringing the virtual in reality, or the other way around, it's the boundaries between them that must be the theme of the movie. Otherwise, I have absolutely no interest in games.
> (www.avalonthemovie.com)

Despite the director's disclaimer, issues of reality vs. virtuality, addressed tirelessly and tediously in the genre, seem to have the potential to capture audiences weaned on video-games and the internet. What interests me in *Avalon*, and has been largely overlooked in discussions of the film, is the setting and production in Poland, with Polish actors and design teams, and the traces of a particular memory of Europe that this setting retains in the genre's cyberpunk imaginary, based as it is on the popular marriage between technological war games and the counterculture.

EASTERN GREEN: TECHNOLOGY OVER NATURE

The monochromatic world of both the game space (with its ochre hues) and Ash's drab everyday reality (in the green-ish tones of early computer-screen technology) are set in Poland, and the »Central European« feel has been noted in reviews of the film (see, for example, Elley 24). Oblique camera angles, crumbling and somewhat threatening cityscapes and other visual markers evoke World War II newsreels and noir movies like the *Third Man*, while the graphic aspects and the rather slow pace (for an action movie) evoke the graphic-novel derived animation style the director is most familiar with. Particularly the painstakingly authentic tanks, uniforms and weapons (pet objects of Japanese *otaku* microcults), which were specifically commissioned from Polish designers, very obviously and vividly evoke a Cold War vision of the pre-89 East bloc (for Western viewers), abstracting dominant images of Europe into game space. This visual style conveys a pervasive atmosphere of (almost) inescapable surveillance and oppression by an authoritarian police state, which, though never explicitly elaborated in the movie, still seems to form the background of Ash's lifeworld. The director makes it very explicit in an interview that this is a deliberate stylistic choice in *Avalon*:

> For Avalon, I had images in my mind of Hungary or Tchecoslovakia [sic], images of civil war […] It's maybe related to my student years in the 70's. At that time manifestations were violent. As a director, every time I make a movie, it relate [sic] to that period. Both in my animated movie and fiction movie, there is a struggle between a repressive police and terrorists. (www.avalon-movie.com)

Eastern Europe serves here as a (mostly visual) trope of a repressive, totalitarian system, and the imaginary of Ash's lifeworld thus contains clearly discernible traces of this part of Europe's (violent and oppressive) history; the World War II allusions to carnage in Poland in the game are juxtaposed with its (Communist) after-effects in Ash's post-45 (and pre-89) experience. At the same time Oshii clearly relates those traces of state violence made in Europe, its history of totalitarianism and warfare, to the equally violent Asian historical imaginary of police oppression and paternalistic or authoritarian social and political structures, with their inclination to label any kind of resistance (often sustained by youth and student movements) »terrorist.« Like his filmic vision, inspired by American cyberpunk yet itself an inspiration for Hollywood, Oshii's political imaginary is *both regional and global* in its evocation of history; setting his story in Poland makes history specific for a certain *period* (in the director's own memory) by making it specific to, and situating it in a particular *place* that retains the traces of, this history.

What makes the juxtaposition of a bleak lifeworld and an ultraviolent gamespace so interesting is the typical, counterculture-driven ambivalence and productivity of the border-crossing between the worlds (something the Wachowski brothers profoundly misunderstood in their adolescent treatment of the same issue). In *Avalon*, it is very clear that the traces of »real« history in the computer game serve as references to the militaristic basis the real world is built on, presenting the players with a constant reminder of the paranoid, defense-related Cold War roots of the computer technology they are using to escape from an equally paranoid society outside the game space (Bell 9; Edwards 12). What unites European images of the (European) »East« before 1989 and the cyberpunk attitude is a fascination with how this »old order« – violent, authoritarian and oppressive – coexists with the virtual imaginaries of its violent roots, and is therefore under assault from being »hacked«: technology being turned against the basis on which it is built. The violence of the game provides at least some of its players with the possibility of agency, whereas the political structures in Ash's real world make them immobile. (When Ash leaves the gaming parlor in an early scene in the movie, she moves among completely static people in the streets – something we are used to seeing in anime, but not in an action film.)

Central Europe is thus a highly apposite location for Oshii's concerns with militarism, technology and surveillance (topics that had already pre-occupied him in his anime *Patlabor 2* (1995) and *Jin Roh: The Wolf Brigade* (1998)), despite his protestations to the contrary. The global cultural memory of the Cold War, maintained in the popular culture genre of computer games on which *Avalon*'s plot and style rests, very economically evokes these themes without specifically referring to any actual contexts, no doubt one of the reasons why it is attractive for a global fan base. The bygone Communist past of Eastern Europe is indelibly inscribed in (almost nostalgic) imaginaries (one cannot evoke Eastern Europe globally without these references), but this evocation becomes a cipher for an »old order« in a wider sense.

YEARNING FOR REDEMPTION: ARTHURIAN LEGEND AND ART MUSIC

Pre-1989 Central Europe is not the only evocation on which Oshii's film hinges. The world of *Avalon* is permeated with references to a fantasy world that has certainly had an impact on computer games and fantasy movies (most recently Antoine Fuqua's 2004 epic *King Arthur*), but, more importantly, can be considered one of the most pervasive and recurring fantasies of the Western imagination: the Arthurian legends, recreated repeatedly with great success (particularly in Mallory's c. 1470 adaptation *Le Morte D'Arthur*, on which not only that canonical work of Anglo erudition, Alfred Lord Tennyson's 1888 *Idylls of the King*, is based, but also John Steinbeck's 1976 posthumously published, popular re-telling *The Acts of King Arthur and His Noble Knights* and John Boorman's 1981 hit movie *Excalibur*). Oshii's references in *Avalon* are well informed and knowing, both in his literary allusions and in the visual style in which some of the more directly Arthurian characters are conveyed (more pre-Raphaelite than medieval); however, this aspect seems to have eluded reviewers of the film. The gritty cyberpunk feel of most of the movie seems to distract from the strong moral message conveyed through the use of fantasy. The director himself does not seem too sure of the importance of this theme himself, but clearly European fantasy, whether in the form of fairy tales or legends, informs his work significantly:

> [Question] You seem to be very influenced and inspired by the European culture. […] You already used tales by the brother [sic] Grimm and Charles Perrault for the script *Jin-Roh*, and now for *Avalon*, the Arthurian legends. How did you get acquainted with these old European stories?
> [MO] I'm often asked why I like these old European legends so much and it's not easy for me to give a clear answer to it. I did like them from very early on, from when I was young. Somehow, I can't really explain why, Japan as an environment to set my movie does not seem to fit my work. Ever since about 20 years ago, when I first became a film director, I didn't really fit in the Japanese film world. Maybe it has something to do with the fact that I am an animation artist, and that I spend a lot of time in imaginary worlds during the creative process. So I don't have great interest in the reality of what Japan is like these days, as a film subject at least. So that is how Europe always seems to come into the picture. But if you ask me why […] I'm not sure that I can really explain. I do have a big interest for Europe in the Celtic age. For *Avalon* as well, the elements of the Arthurian legends represented something I really wanted to use. I often talk about this with this other famous manga director, Hayao Miyasaki (director of *Princess Mononoke*). The European influence is psychologically very complex, it is not just that we are simply infatuated by Europe, it is more that we really dislike certain aspects of the way things are in Japan these days. And of course in animation it is easy to situate everything in a different world which is not Japan. If you look at Japanese animation films these days, most of them won't have characters with Japanese faces. It is not that they have any other nationality in particular, but they are definitely not Japanese. And also for Avalon, of course, I really liked the general impression I had of polish faces when we went looking for location. Again it's not easy to explain why […] but it isn't just me: there seem to be quite a lot of people who agree with the fact that they have a great dissatisfaction with the reality of Japan these days. That's why we shift our focus towards Europe. (www.avalon-movie.com)

Obviously it is not only European history but also European *culture* – in this case Arthurian mythology – which serves as a global frame of reference for making a political statement that could be understood in a wider context, at least by the global art-house crowd. Often overlooked in discussions about the spread of *American* culture under the conditions of cultural globalization, centuries of spreading an imperialist concept of European *Bildung* have clearly resulted in a pervasiveness of narrative sediments in popular culture, and Oshii is able to mobilize them very elegantly in his movies. In *Patlabor 1* (1990) it is the biblical tale of the Babylonian tower that provides a frame of reference for a story about a computer virus, unleashed because of the master criminal's frustration over the environmental destruction of Tokyo; in *Jin Roh: The Wolf Brigade* (1998) it is the Little Red Riding Hood motif (and story) that frames a narrative about betrayal and the impossibility of human trust in totalitarian social structures. In each of these examples, the cultural references are by no means random trinkets, but knowing allusions that add layers of meaning for savvy audiences. In Oshii's play with the notion of »what is real« under the impact of technology-mediated virtualities, the European references are usually situated very cleverly on both sides of the fence (in viruality as well as reality): the computer virus in *Patlabor 1* ostensibly exists in cyberspace, but it has an influence on robots which go on a rampage in the real world, destroying the highrises and bridges that signify the contemporary city. The *Rotkäppchen*-metaphor in *Jin Roh* (*Rotkäppchen* is also the name of the German WWII Ruhrstahl/Kramer X-7 missile, something Oshii and some of his fan base would know and which goes to prove that even fairytales retain memories of Europe's violent history) is employed in the murky realm between reality and dream, in a case of mistaken identities, misrecognition and deception. (In the same movie, a reference to Gottfried von Strassburg's *Tristan und Isolde* prefigures the Arthurian motifs in *Avalon*.)

It is thus by no means surprising that Oshii uses Arthurian legend in *Avalon* as a metaphor for what his players aspire to. Early in the movie, the notion of *Avalon* is clearly used to point to something beyond *both* the »unreal« shoot'em-up universe and the drab reality of Ash's (and the other players') existence. Interestingly, it is pseudo-art music which gives us the first clues that there is something *beyond* the mutually dependant real/unreal couplet of Ash's life, in a scene where Ash rides a streetcar through the drab streets of her real world while the theme music plays in the background. Again it seems that these are very deliberate evocations of European *culture*, but in this case the music is the first step into a fantasy space in *Avalon* that goes beyond class A – and thus transcends the mutual formation of war game and oppressive reality. But it is a fantasy space, and when we are confronted with it for the first time, it is in a musical style which clearly does not just *imitate* European art music, but widens the frame of reference by creating a *fantasy* of European music for a global audience:

> To emphasize the mythological aspect of *Avalon*, Kenji Kawai has composed a voluptuous »classic« music that was performed by the Warsaw Philharmonic Symphony Orchestra and soprano soloist Elzbieta Towarnicka. For the Japanese Kenji Kawai, capturing the east-european feeling required for the atmosphere of the movie was a real challenge: »At first, I asked myself the question: what is a European sound? Because the European way of pronouncing words, of singing and the methods of composition are totally different from ours.« […] Following the methods used by Mamoru Oshii to modify the original shootings, Kenji Kawai has transformed the original recording of the Orchestra: »Once the orchestra was recorded, everything was transferred in the digital domain. All the parts of the music were cut and set in loops to give them a very systematic regularity. Usually, in an orchestral performance, there are always ›human‹ time shifts. I eliminated them. I transformed this »human« material to give it a mathematical pace.« (www.avalon-movie.com)

The music is clearly as much a technologically enhanced version of what signifies as European as the visual style of the game. But it also serves as an indication of what the mystical final level of the game space indicates: a (technologically based) transcendence of the historical Janus-face of Eastern Europe as violent and oppressive, a way out that can only be found by the fusion between culture and digital wizardry – in game terms by combining the solving of riddles with the fast reflexes of the shooter. This final platform (the one Ash strives to reach for most of the movie) is elevated to a pseudo-metaphysical level: the resting place of Arthur, »the once and future king« (a pseudo-metaphysics which is again cleverly articulating the European *Bildung* myth to the more esoteric imaginary of anime, most famously embodied in Katsuhiro Ôtomo's *Akira* (1988), which also features the »return« of a redeemer, who, not unlike the once and future king, waits for his awakening at an opportune moment.)

But this redemption, the mystical realm which Ash guesses and shoots her way into, also bears the traces of a (more recent) European history, and it is in the final section (or platform) that we find the most compelling problematic of Oshii's political vision.

CLASS REAL; OR: WELCOME TO THE UNION?

The conclusion of *Avalon* takes us to »Class Real,« a level that has almost »end of history« qualities in Fukuyama's sense. After shooting her last enemy (a pale pre-Raffaelite ghost child), Ash wakes up in what appears to be her own quarters, stripped of all her martial warrior-paraphernalia and only in her underwear. It turns out, however, that this is »class special A,« the final platform. She is informed by the game master that there is only one thing for her to do here: »Finish off the unreturned.« All she has for this task is a tiny gun, one clip of ammunition, and a black designer dress in the closet. There is no time limit, and the only exit from this stage, or so she is told, is completion. A poster on the wall features her dog (an obvious link between the worlds) and a concert announcement.

When Ash exits the room she woke up in, Oshii has a neat visual surprise for the viewers that immediately makes it clear what »Special A« is all about. While the movie has up to this point (over an hour) accustomed the audience to a greenish and sepia monochromatic world and a rather slow pace (even the shoot'em-up scenes are in slow motion), all of her sudden Ash finds herself in the street of a colorful modern metropolis, complete with brand ads that feature Western consumer culture and lifestyle. The first visual impression we get of the street is a sunlit Eastern European façade and a yellow streetcar with a deep-blue Nivea ad; the camera then cuts to a bright-red Coca-Cola ad and another highly colorful poster, presumably a cigarette ad, that says »Taste It – the West.« Gone are the weapons, the armor and the shootouts (as well as the elegiac slow-mo of Ash's humdrum native city); Ash fits right in with her designer dress – her warrior outfit would have been most gauche in what is obviously an affluent post-89 market economy in an Eastern European city (Warsaw), and not only the streetcar, but also people move fast, faster than Ash, who moves hesitantly in this new environment (a significant contrast to the earlier scene in which she is the one passing completely immobile people on a bridge in her previous »real world.«)

Oshii's acclaimed visual coherence in the first three quarters of the movie obviously sets us up for this stunning, and highly significant, visual break. Viewers would at this point expect to be treated to more profundities about the status of the real at this final level of the movie, and Oshii does not disappoint, employing the shopwindow metaphor to excess in the first few minutes of Ash being in her new environment (we see Ash observing a street which is reflected in the window of a department store – not exactly a subtle point). But the European (and again global) relevance of these scenes is loud and clear, and again addresses the relevant political dimension of this Brave New World (a world without an exit strategy?).

Avalon, the mystical realm of redemption, is (virtually) imagined in the form of the (colorful) trappings of consumer culture, overlaid with what has signified as »Western« tradition in the form of art music culture. The bleak world of authoritarianism seems to have been superseded by commodity culture, an obvious parable of the state of Eastern Europe today after joining the EU, which seamlessly incorporates traditional *Bildung* without breaking its (fast-paced) stride. American reviewers have had a field day with this obvious reading, a clear sign that the Cold War is still a major factor in the reception of any cultural product which deals with Europe:

> Ash has almost no life outside Avalon. Nor, in a sense, does Avalon. The game represents a yearning for some sort of redemption (as a choral group sings, »Avalon, when will that day come?«). The movie offers it – somewhere beyond anime. Oshii has admitted that using a Polish cast gave him license to digitally manipulate their facial expressions in post-production. Special A is supposed to be »realer than real«; and in the movie's final movement, Ash makes a breakthrough from her drab, near-monochrome

> environment into a new world of color and advertising. Among other things, *Avalon* may be the first movie that uses contemporary Poland as a special effect. (Hoberman)

Contemporary Poland as a special effect is a nice tag line, but Avalon signifies more than a special effect in Oshii's imaginary. Ash's bemused wandering through the streets of present-day Warsaw confronts her with every symbol of modern urbanity: cars, streetcars and subways, and people moving deliberately and with purpose. Even the basset is chauffeured around in a car, enjoying city life like an affluent *flâneur*, so surely there is more to Special A than colorful effects.

The lead of the dog brings Ash to a concert matinee, and it is here – again in the context of art music being performed in a historic (albeit reconstructed) concert hall – that she encounters the target of her final mission: the »unreturned« Murphy. Predictably, they have a long conversation about reality in the park-like courtyard of the concert hall (framed by old cannons) while the theme music is performed on stage. Murphy admits that he set the game up deliberately so it would lead to his brain death and ascension to Class Real before he forces Ash into a confrontation and gets fragged. It turns out that this was a form of virtual suicide: he had no bullets in his gun. But there is no »mission complete« signal – Ash stays in this world, encounters the ghostly girl again, points her gun at her, and the movie ends with a computer screen that says »Welcome to Avalon.«

This ending is deliberately ambiguous; no pseudo-religious redemption à la *Matrix Revolutions* here. According to Oshii, »[a]ll films are inconclusive. Just like things called ›life‹ and ›reality‹« (e-mail, quoted in Ruh 183). Ruh speculates that the Arthurian reference might imply that Ash is supposed to return to her world as a redeemer (bringing Nivea ads to the crumbling city?):

> Class Real is not a true reality, nor is it the ideal world for Ash. Rather, Class Real is another illusion, albeit a seemingly welcome one in which she may have power and immortality. […] Ash must strive to wake from an imposed dream that is not her own. Ash recognizes the illusion of the world and makes a conscious decision to leave Class Real. (Ruh 180)

I find this interpretation highly unconvincing. As with the destiny of the cyborg Kusanagi in *Ghost in the Shell*, there does not seem to be a way back for Ash; Major Kusanagi has to stay in cyberspace, and Ash in the oneiric consumer culture of EU-Eastern Europe. Despite the visual break, there are many connections between this Class Real and previous levels of the game; despite the art music, the well-dressed people and the classy, genteel and well-groomed location, there are still traces of a violent history (the cannons), and there is still the need to assert oneself violently and with very little material help – after all, Ash comes into this level stripped of all her previously acquired items. (The lyrics of the theme music say »so many of the brave are cut down and heroes lie fallen« – Arthurian legend is not exactly pacifist either.) The dog might have advanced from real vegetables to being driven around in a sedan, but Ash, leaning against a cannon, still has to pick up the bullets and reload after getting rid of Murphy.

Whatever Ash's future holds (we might learn more about it if there is to be an *Avalon 2*, similar to Kusanagi's fate in *Ghost in the Shell 2: Innocence*), the political parable Oshii lays out for us again cleverly employs images of Europe to make a point about global political phenomena. The new »class real,« the supposed redemption from authoritarian and paternalist structures by a globalized culture of brand names and feverish status consumption (including, yes, Europe as a label and art music as a style) might reflect the current reality in Eastern Europe, but it is a global phenomenon that has had an equally violent impact on Asia. If we look at the recent histories of democratic countries in the region which have recently made the shift from strictly authoritarian regimes (Taiwan, Korea), or at what is going on in China right now, the grail certainly has a price tag, and global capitalism's game world defines most people's everyday reality, a reality born out of violence yet commercialized, just like the cyberspace universe that is its enabling condition – in other words, a real-world fantasy come true, without an obvious way out or »reset« button.

What Oshii seems to indicate in the final scenes is that although the rules might have changed (no heavy weapons, but also no time limit), this world is different, more colorful, but certainly no less violent than the levels »below« it – the levels on which Class Real is built –, and its ontological status remains uncertain. Yet, Oshii seems to suggest, there is no way back, easy or otherwise. The visual break indicates clearly that whatever came previously has to be relegated to the realm of nostalgia. In the meantime, the loops of the digitally enhanced, European-branded art music keep playing globally.

REFERENCES

Bell, D. *An Introduction to Cybercultures*, London; New York, N.Y: Routledge, 2001.

Edwards, P.N. *The Closed World: Computers and the Politics of Discourse in Cold War America.* Cambridge, Mass.: MIT Press, 1996.

Elley, D. Avalon. *Variety* (2001): 24.

Hoberman, J. Avalon. *The Village Voice*, 31 Dec. 2003–6 Jan. 2004, C55.

Leggat, G. »Also of interest.« *Film Comment* 40, 1 (2004): 79.

Ruh, B. *Stray Dog of Anime: The Films of Mamoru Oshii.* New York: Palgrave Macmillan, 2004.

Sobchack, V. *Screening Space: The American Science Fiction Film.* New Brunswick, N.J.: Rutgers University Press, 1997.

Contributors

Christina Baade, Assistant Professor, Music and Communication Studies, McMaster University, Hamilton, Canada.

James Deaville, Associate Professor and Director of Graduate Studies, School of the Arts, McMaster University, Hamilton, Canada.

Susan Ingram, Assistant Professor, European Studies, Humanities, York University, Toronto, Canada.

Srdja Pavlovic, Assistant Adjunct Professor, Department of History and Classics, University of Alberta, Edmonton, Canada.

Markus Reisenleitner, Head, Department of Cultural Studies, Lingnan University, Hong Kong.

Michael Saffle, Professor of Music and Humanities, Center for Interdisciplinary Studies, Virginia Tech, Blacksburg, U.S.A.

Peter Stachel, Österreichische Akademie der Wissenschaften, Kommission für Kulturwissenschaften und Theatergeschichte, Vienna, Austria.

Cornelia Szabó-Knotik, Head, Department of Music Theory, Analysis and History, University of Music and Performing Arts, Vienna, Austria.

John C. Tibbetts, Associate Professor of Theatre and Film, University of Kansas, Kansas City, U.S.A.

Takenaka Toru, Graduate School of Letters, Osaka University, Osaka, Japan.

Simon Wood, Music and Multimedia, School of the Arts, McMaster University, Hamilton, Canada.

Andriy Zayarnyuk, Research Fellow, Institute of Historical Research, Ivan Franko Lviv National University, Lviv, Ukraine.

Peter Lang · Europäischer Verlag der Wissenschaften

Susan Ingram / Markus Reisenleitner /
Cornelia Szabó-Knotik (Hrsg.)

Ports of Call

Central European and North American Culture/s in Motion

Frankfurt am Main, Berlin, Bern, Bruxelles, New York, Oxford, Wien, 2003.
XIV, 283 pp., num. fig., tab. and graphs
ISBN 3-631-50696-1 / US-ISBN 0-8204-6416-3 · pb. € 51.50*

This volume collects papers put together by an interdisciplinary group of scholars, which explore how the two imaginary geo-cultural spaces "Central Europe" and "(North) America" have mutually attributed meanings to each other over the past two centuries, how traveling images of an »othered« cultural space – inserted into specific regional, national and social contexts and appropriated for negotiations of cultural identity and belonging as well as exclusion and colonization – have laid the basis for a cultural essentialism which thinks culture through space and negotiates cultural status through de-historicized notions of place and territory. It particularly focuses on processes of motion and travel which helped to create these images and discusses in individual case studies a wide variety of cultural phenomena – ranging from music to film, from tourism to world fairs – while sharing the common concern to explore how motion through space – whether physical or imaginary – helped shape, crystallize and negotiate images of the cultural other in contact or transit zones where people, images and cultures meet in asymmetrical relations of domination and subordination, and where tourists, exiles, travelers, displaced commodities and foreign cultural practices generate powerful, as well as potentially subversive, visions and imaginings. Thus this volume invites to find individual paths and ports in/between the subjects presented and in a way to contribute to, to follow up the web of exchange represented by its authors, themselves a (mostly) virtual community of researchers.

Contents: Studies on traveling images of an "othered" cultural space in music, film, people and sites · Callers and Callings · Traveling Sites, Sounds and Visions

Frankfurt am Main · Berlin · Bern · Bruxelles · New York · Oxford · Wien
Distribution: Verlag Peter Lang AG
Moosstr. 1, CH-2542 Pieterlen
Telefax 00 41 (0) 32 / 376 17 27